Post-Parametric
AUTOMATION
IN
DESIGN AND CONSTRUCTION

Post-Parametric
AUTOMATION
IN
DESIGN AND CONSTRUCTION

Alfredo Andia
Thomas Spiegelhalter

Library of Congress Cataloging-in-Publication Data
A catalog record for this book is available from the U.S. Library of Congress.

British Library Cataloguing in Publication Data
A catalog record for this book is available from the British Library.
ISBN-13: 978-1-60807-693-2

Cover design by John Gomes

Cover image courtesy of RMIT Architecture Masters Studio, 100 YC: Tom Kovac, Michael Mei Students: Wencheng John Xu, Miau Teng Tan, Dac Thanh Vu

© 2015 Artech House

All rights reserved. Printed and bound in the United States of America. No part of this book may be reproduced or utilized in any form or by any means, electronic or mechanical, including photocopying, recording, or by any information storage and retrieval system, without permission in writing from the publisher.

All terms mentioned in this book that are known to be trademarks or service marks have been appropriately capitalized. Artech House cannot attest to the accuracy of this information. Use of a term in this book should not be regarded as affecting the validity of any trademark or service mark.

10 9 8 7 6 5 4 3 2 1

Contents

Preface

Post-Parametric Automation in Design and Construction   13
(Alfredo Andia and Thomas Spiegelhalter)

Automating Design   13

Automating Construction   14

Toward a Semiautomated Construction Sector   14

Part I—Automating What?

17

Chapter 1

Toward Automating Design and Construction   19
(Alfredo Andia and Thomas Spiegelhalter)

Introduction   19

What Are Computers and How Do We Use Them?   19

Automation   20

Automating Design vs. Automating Construction   20

The Automation Themes in Architecture and Engineering: From CAD to Parametric   21

Brief History of Parametric in Architecture   21

Three Parametric Paradigms   22

Post-Parametric Era   23

Automating Architecture and Engineering via Machine Learning   23

Automation Themes in Construction   23

Automating Construction via the Future of Digital Manufacturing   24

Conclusion   25

Chapter 2

Green Automation: Design Optimization, Manufacturing, and Life-Cycle Sustainability   27
(Alfredo Andia and Thomas Spiegelhalter)

Introduction   27

Toward Interoperable, Automated, Parametric/Algorithmic Carbon-Neutral Design Workflows   28

Total Green Building Automation System with Human-Computer-Interface Topologies   30

Automation in Green Building Manufacturing   31

Conclusion   32

Part II—Post-Parametric Workflows in Architectural and Engineering Offices

(Alfredo Andia and Thomas Spiegelhalter)   35

Chapter 3

Engaging with Complexity: Computational Algorithms in Architecture and Urban Design   39
(Keith Besserud, SOM)

Introduction	39
Search Algorithms	40
Genetic Algorithms	41
Systems Modeling	42
LakeSIM: Integrated Infrastructure Modeling Platform	44
Conclusion	45

Chapter 4

Space Planning with Synthetic User Experience   47
(Christian Derix, AEDAS)

Introduction	47
Space as Heuristic Organization	48
Models of Designer as User	49
Models of Occupant as User	50
Relational Representation	50
Enactive Architectures	51
Generic Functions of Buildings	53
Designing Organically	55
Computational Archetypes of Space	56

Chapter 5

Algorithmic Principles for Façade and Building Automation Systems: Al-Bahar Towers, Abu Dhabi   59
(Abdulmajid Karanouh)

Introduction	59
Key Design Elements	60
Adaptive Mashrabiya Solar Screen	61
Managing Complexity	62
The Algorithm	65
Setting Up First Set of Algorithmic Principles and 4-D Java Scripting	66
4-D Parametric/BIM Model and Geometry Optimization	70
Adaptive Principles Optimization, Construction, and Performance Manual	72
Updating the Adaptive Algorithmic Principles and HMI Control Software	74

Chapter 6

Custom-Designed Structures and Façades with Parametric-Algorithmic BIM Systems: 1 Bligh Street, Green Star Rated High Rise Project, Sydney — 75
(Thomas Spiegelhalter)

Introduction	75
Site Condition, Building Key Features, and Systems	76
Double Façade, Space Conditioning, Cooling, and Energy Use Concept	79
Challenges of the Multidisciplinary CAD To BIM Collaboration	79
Structural Analysis and Design Integration	83
Integrated Double Façade Performance Analysis, Mechanical, Electrical, Plumbing, and Fire Service Design	84
Interoperability with the Contractor and Subcontractors	84
Renewable Energy, Water Recycling, and Benchmarking	85
Conclusion	86

Chapter 7

Parametric-Algorithmic Automated Modeling and Fabrication: The Railway Station Stuttgart 21 — 89
(Albert Schuster, Lucio Blandini, and Thomas Spiegelhalter)

Urban Large-Scale Project Main Station Stuttgart 21	89
Parametric-Algorithmic Design of the New Railway Station Stuttgart 21	90
Nonlinear Analysis and Structural Behavior Optimization	92
Finite Element 3-D Modeling and Automation	93
Scripting and Fabrication Process	95
Assembly Process	96
Low Primary Energy Requirements	96
Zero-Energy Station and Passenger Comfort	97

Chapter 8

Integrated Project Delivery and Total Building Automation: Q1 ThyssenKrupp Headquarter, JSWD Architekten + Chaix & Morel et Associés — 99
(Thomas Spiegelhalter)

Introduction	99
Q1 Sun and Daylight Prototype Control System: Design and Production	101
Q1 Energy Concept	107
General Q1 Project Management	107
Q1 Total Green Building Automation (DESIGO)	109
Q1 Certifications, Awards, and Honors	109

Chapter 9

Design Computation at Arup   112
(Clayton Binkley, Paul Jeffries, and Mathew Vola)

Introduction	112
Parametric Design Case Study	114
Custom Tool Development	116
DesignLink SDK	116
SALAMANDER	116
Stadium Generator	117
Automation	118

Chapter 10

Generic Optimization Algorithms for Building Energy   121
Demand Optimization: Concept 2226, Austria
(Lars Junghans)

Introduction	121
Optimization Methods: Probabilistic Optimization Methods	122
Optimization Methods: Sequential Search Algorithms	124
Available Optimization Tools	127
Future Developments in Generic Building Optimization	127
Built Example	128
Conclusion	129

Chapter 11

Customized Algorithmic Engineering of a Curved Cable -   131
Stayed Façade: The Enzo Ferrari Museum, Modena, Italy
(Lucio Blandini and Werner Sobek)

Introduction	131
Geometry	133
Cable-Stayed Glass Façade	134
Aluminum Roof	138
Conclusion	139
Acknowledgments	139

Part III—Post-Parametric Automation in Construction

(Alfredo Andia and Thomas Spiegelhalter)   141

Chapter 12

Siemens Digital (Self-Learning) Factories and Automation: 145
Automated System Optimization via Genetic Algorithms
(Thomas Spiegelhalter)

Traditional 2-D Factory Design Processes Are Prone to Error	145
Digital Factory Design and Operation with PLM Software	146
Case Study: Integrated Tecnomatix and Robotics Process Engineering for the Volkswagen Group	149

GAs, Neural Networks, and Wasp Swarm Optimization of Logistic Systems and Automation	151

Chapter 13

Prefabricating a More Sustainable Building and Assembling It in 15 Days: Broad Group, China — 155
(Alfredo Andia)

The Origins	155
Sustainability Vision	156
Broad Sustainable Building	156
The T30 Hotel Built in 15 Days	157
BSB Sustainability	158
Cost and Time	158
Building Automation System	161
Business Model	161
Sky City: The Tallest Building in the World Built in 90 Days	161
Conclusion	161

Chapter 14

Automated Fabrication and Assembly: Sekisui Heim, Tokyo, Japan — 163
(Jun Furuse, Masayuki Katano, and Thomas Spiegelhalter)

Introduction	163
Modular Sekisui Unit House	164
Workflow from Client's Design Contract to Manufacturing with Robots to On-Site Assembly	165
Technical Key Points In The Arrangement and Programming of Parts	166
BOM Structuring	166
Parts Arrangement System and Outline of HAPPS	167
Conversion of Intermediates to Objects	168
Property Inheritance from Object to Parts	169
Application Scope of HAPPS Information	169
Summary of the Efficiency and Accuracy of HAPPS	169

Chapter 15

Customized Prefabrication in Two Hospitals: NBBJ, Ohio — 171
(Alfredo Andia)

Introduction	171
Miami Valley Hospital: Implementing the Idea of Prefabrication	171
Prefabrication Performance Metrics	176
Neuroscience Institute at Riverside Methodist Hospital	176
Colocating of the Entire Design/Build Team in NBBJ Office	177
Just-in-Time Prefab Construction Schedule	177
3-D and 4-D BIM Models	178

Improvements in the Prefabrication of the Components	179
Conclusion	179

Chapter 16

Robotic Fabrication: ICD/ITKE Research Pavilion 2012 — 181
(Achim Menges and Jan Knippers)

Introduction	181
Biological Model	182
Transfer of Biomimetic Design Principles	183
Computational Design and Robotic Production	184

Part IV—Emerging Automations

Chapter 17

Automating Design via Machine Learning Algorithms — 191
(Alfredo Andia)

Introduction	191
Limitations of Parametric Systems	191
Algorithms vs. Learning Algorithms	192
Computers as Autopoietic, Self-Organizing, and Self-Learning Systems	192
Parametric: First Stage of AI	193
Machine Learning: Second Stage of AI	193
Examples of Machine Learning Algorithms Outside the AEC industry	193
Learning Algorithms in Architectural Design	194
Automated Design for Residential Building	194
Automating Building Layout Design	195
Machine Learning Hardware: Neuromorphic Processors	197
General AI: Third Stage of AI	198
Conclusion	199

Chapter 18

Automating Construction via n-D Digital Manufacturing — 201
(Alfredo Andia)

Introduction	201
At the Dawn of Three New Manufacturing Eras	201
3-D Manufacturing: Making Any Form in 3-D	202
3-D Manufacturing: Large-Scale Digital Manufacturing	202
4-D Manufacturing: Printing New Materials	203
4-D Manufacturing: Adaptive Materials	205
4-D Manufacturing: Nanotechnology	205
n-D Manufacturing: Programmable Matter	205
n-D Manufacturing: Self-Made Robots	206

n-D Manufacturing: Synthetic Biology — 206
Conclusion — 206

Chapter 19

Conclusion: Another Look at Semiautomating the AEC Sector — 209
(Alfredo Andia and Thomas Spiegelhalter)

1910s–1930s: Explosive Industrialization — 209
1950s: Standardized Industrialization — 209
2010s: Explosive Digital Innovation — 209
Who Owns Innovation Outside the AEC Industry? — 209
Platforms of Digital Innovation — 210
AEC Social Units — 210
Digitally Disrupting Platforms in the AEC Industry — 210
Machine Learning: Automating Design — 211
Big Data: Automating Planning and Real Estate Development — 211
Digital Manufacturing: Automating Construction — 212
Examples of Emerging Digital Manufacturing: Robotics — 212
Examples of Emerging Digital Manufacturing: 3-D Printing — 212
Routes of Digital Consumption — 213
Cautionary Tale — 214
Moonshot Thinking — 215

About the Editors — 217

About the Authors — 219

Index — 223

Preface
Post-Parametric Automation in Design and Construction

Alfredo Andia and Thomas Spiegelhalter

This book is not only about design and technology but it is about the automation narratives innovative social units are developing for the construction sector. Automation, a mixture of algorithms, robots, software, and avatars are transforming all types of jobs and industries. Algorithms today have automated around 70% of trades in U.S. stock markets, define the patrol route for the Los Angeles police, write news without human intervention, allow cars to drive autonomously, beat human champions in the TV show *Jeopardy!*, and select companies that receive venture capital investment in less than two weeks. Robotic apparatuses fulfill orders in the vast Amazon warehouses, fold clothes at a Berkeley lab, and accurately slice meat for supermarkets. Will automation impact the design and construction industry?

This book is organized around how architectural, engineering, and construction (AEC) professionals are developing their automation narratives. We argue that there are two types of major automation discourses today: one that emphasizes the automation of design and another one that searches for automation from the perspective of how buildings are constructed.

In Part II of this book we look at how technologically advanced architectural and engineering practices are semiautomating their design processes by using sophisticated algorithms to transform their workflows. In Part III we document a set of firms that are further advancing automation by using prefabrication, modularization, and custom design via robotics. In Part IV we look at the future, and we argue that there will be two different forces that will further automate the construction industry: machine learning and digital manufacturing—both of which will evolve rapidly in the next decade.

Automating Design

In the next decade automation will move the subject of computers in design way beyond the computer graphic narratives (computer-aided design (CAD)/building information modeling (BIM)/parametric) that have dominated architecture and engineering in the past two decades. The computer graphics paradigms that have haunted architecture and engineering in the past two decades were very much related to the old software paradigms that matured at Xerox Park in the 1970s and were popularized in the 1980s with the emergence of the personal computer.

Software metaphors such as computer graphics and parametric systems are considered in design theory of computer sciences as the most primitive stage of artificial intelligence.

Today, we are well into a second era of artificial intelligence (AI) in which algorithms can learn from data without the assistance of a human. A whole generation of diverse products from Internet search, automated translation, forecasting energy consumption, managing energy, vehicular traffic estimation, drug design, and fraud detection are the result of learning algorithms. With learning algorithms we are moving away from manually coding systems to designing systems that learn from experience. We are in the first steps of creating sophisticated machine learning algorithms that develop specific intelligence in design synthesis, building simulation, operation, control, and benchmarking.

Automating Construction

Two parallel discourses of automation are emerging from the construction point of view. Contractors are moving into manufactured prefabrication, and architects and engineers are advocating for custom fabrication. On the one hand we present how some major contractor prefabricators have shifted the majority of labor hours from the construction site into highly sophisticated facilities to significantly reduce costs, materials, schedule, and the environmental impact of construction. In China the Broad Group has built a large number of modular construction structures, including a 30-story hotel that could be prefabricated in a factory in 7 days and be assembled on site in 15, which significantly reduces the carbon footprint and lowers the cost of construction to $50 per square feet ($500 per square meter). On the other hand we present how architects and engineers are developing custom fabrication and mass customization techniques by developing in one case a large number of subassembly units just-in-time and in another case using robotics to achieve very unique design performance and life-cycle design quality.

We think that these advances in the profession will be further challenged by the acceleration of digital manufacturing platforms. The only certainty about digital manufacturing processes is that they will not disappear but on the contrary they will grow exponentially. Today we have experienced only 3-D digital manufacturing platforms such as 3-D printing, computer numerical control (CNC), laser cutting, and inflexible robotics. 3-D digital manufacturing alters materials on a large scale. We are quickly entering into a 4-D digital manufacturing period in which we will be able to design, create, and print all sorts of new environmentally sound materials at a microscopic level, inventing materials that cannot be found in nature. In a longer horizon we will began to see the emergence of an N-D digital manufacturing era in which materials can be programmed and be malleable at will.

Toward a Semiautomated Construction Sector

This book assesses the current status of automation in the design and construction industries and critically evaluates new forms of practice and processes. The story of this rising era of automation is not just a technological and environmental one, but it is a highly social-cultural one. In contrast to industrialization, automation is not a standardizing technology but on the contrary it is allowing social units to emerge with very precise local themes, which afford customization that targets very precise type of endeavors. Automation is not technology but the construction of the organizations that conceive it. The digitalization of construction will come in a series of steps, platforms, and innovative social processes. We think we are far from fully automating the construction sector, but we are definitely entering into a period in which we are semiautomating a significant number of tasks that will lay the foundation to transform our analog world.

In the end we question which social units will be in charge of these changes. We also question how much of the architecture, engineering, and construction (AEC) trade will became an information technology business as is occurring in many other professions.

Part I

Automating What?

Chapter 1

Toward Automating Design and Construction

Alfredo Andia and Thomas Spiegelhalter

Introduction

This book responds to one critical question in the design and construction industry these days: How are architects, engineers, and contractors using information technology to further automate their practices? Automation in this book implies important saving themes in labor, energy, materials, and construction quality; and significant improvements in sustainability to meet new international standards of carbon neutrality.

The term automation began to be popularized by the American car industry in the 1940s and 1950s. The term was used to describe automated mechanization that was maturing in all types of production lines at the time. The term automation began to change in the middle of the 20th century with the introduction of tools such as numerical control (NC) machines that were automatically controlled by coded mathematical information saved in punched cards.

Today algorithms, scripting, robots, digital manufacturing, and new autonomous workflow systems are once more transforming the meaning of the term automation. A new level of digitally based automation control over production, services, and even social media continues to surprise and constantly transform as exponential growth of cheap computing power prolongs its course.

What Are Computers and How Do We Use Them?

In order to comprehend how architects are adapting and using digital technology, we must first address two key questions: What are computers? and How do we use computers? Computers, as invented by Charles Babbage and Ada Lovelace in the 19th century, are a particular type of machine: an all-purpose machine. Thus, the imagery, the charisma, and the themes of computerization are and will be constantly shifting and adapting to new types of imaginations—each time at a faster pace [1].

How do we use these all-purpose machines? Computerization is more than a technological phenomena; it is a consumer phenomena. Computers are consumed in a social context. We use computers to talk about our visions about the processes, organization, and culture of our disciplines. Even though the media often treats computerization shifts as revolutionary, most of the computerization themes developed by professionals or managers are relatively simple and usually are intended to impact only the social unit or the narrow context in which the organization operates. For example, it is difficult to find an architectural firm that is imagining the automation of construction processes. Vice versa, it is not easy to encounter contractors that are automating architecture and engineering processes. Thus, there are many parallel narratives of automation across social units and disciplines.

Automation

It is important here to place in context the term automation. Detroit automation of the mid-20th century was particularly important for shaping the contemporary image of automation. As critical historian David F. Noble and management consultant Peter Drucker suggested, Detroit management was interested in using mechanically and digitally automated equipment to continue Taylor's techniques of subdividing tasks that could be ultimately performed by machines rather than by humans [2]. This further implies that the discourses of automation are shaped by the social conversations and the management goals rather than pure technological determinism.

Today the word automation is usually associated with digital manufacturing processes found in the aerospace, aeronautical, ship building, and automobile industries. These industries have a high level of automation, but they also have a very different social organization and funding structure than the architecture, engineering, and construction (AEC) industry. For example, an average new car plant can cost approximately 1 billion dollars, and a single car product can easily surpass 100,000 hours of engineering. In comparison, the AEC industry can only invest a much smaller number of professional hours to produce a much larger product that has to adapt to stringent local regulations, wider customer choice, and an array of site conditions [3]. The differences between processes used in car manufacturing and housing manufacturing have been studied in works such as [4, 5].

Automating Design vs. Automating Construction

The AEC industry is a very fragmented industry and is organized around a large number of relatively small social groups that often tend to imagine information technology only within the context of their disciplines and organizational units. Moreover, social imaginations of technology in the AEC industry are limited due to budget constraints as technological investigations often have to be funded as part of specific projects in their professional practice. All this creates a very different type of technological consumption phenomenon that has a noteworthy dependence on the vision of software vendors. Designers and contractors have developed two major divergent automation narratives today.

1. *Automation themes in architecture and engineering social units:* A number of architectural and engineering firms are altering their practices by readapting their workflows with parametric, algorithms, building information modeling (BIM), design computation techniques, and scripting tools that help them automate parts of their design, specifications, and fabrication processes.

2. *Automation themes in construction social units:* On the other hand, contractors have begun to transform their practices by moving gradually into more sophisticated processes of prefabrication, modularization, and semiautomated manufacturing.

Parts II and III are organized around the divergent automation narratives that designers and contractors are having today.

Figure 1.1 Automation themes in the AEC industry are often associated to the social imaginations of practice. The images above show the automated precut of the timber frame for a custom made beach house which was assembled on site in one day and designed by the firm Bakoko in Japan. The method, which is widely used in Japan, uses robotic machinery that can cut wood joints following Japanese traditional intricate carved joinery and customary assembly methods. (Images courtesy of Alastair Townsend.)

The Automation Themes in Architecture and Engineering: From CAD to Parametric

In the 1990s design and engineering companies in the developed world were implementing small computerization themes by introducing software such as computer-aided design (CAD) and enterprise programs on personal computers (PCs). However, PC technology only affected skill/manual labor [6]. From the early 2000s the possibilities of doing small automation routines that can script design workflows have moved into the forefront. Some architects and engineers began to use parametric software and scripting to develop parametric design processes.

The most basic conceptualization of parametric refers to a 3-D digital model or digital environment associated with knowledge structures, information, performance properties, and automatic procedures that can aid the designer to construct quick scenarios during design. These models can be updated over time through the Cloud and reused.

Brief History of Parametric in Architecture

Parametric is not new. Parametric ideas in design modeling were an essential feature of the first CAD program, Sketchpad, developed by Ivan Sutherland in 1962. Parametric was also part of the pioneering CAD systems in the early 1970s such as SSHA, CEDAR, HARNESS, and OXSYS. These CAD systems had particular parametric features that were associated to a particular type of knowledge base to serve particular organizations and building types [7]. OXSYS was the precursor of building design system (BDS) and really usable computer-aided production system (RUCAPS), which became available commercially in the UK in the 1970s and surfaced with concepts very similar to today's BIM systems.

All these systems had a common vision: to construct virtually a 3-D building by modeling all their building elements and assemblies. They allowed multi-users to manipulate a single parametric 3-D model in which graphic reports and 2-D drawings were mere automatic derivatives created from the main 3-D model. By the mid-1980s a second wave of 3-D parametrically based software, such as SONATA, Reflex, CHEOPS, GDS, CATIA, GE/CALMA, Pro/Engineer, Solid Works, and many others, achieved a commercial presence. Many of these pioneering parametric programs in the 1980s became standard in industries such as electronics, infrastructures, aerospace, naval engineering, and car manufacturing. However, most practices in the AEC industry preferred to implement 2-D CAD systems in PCs. It took close to two decades for the 3-D parametric model to make a significant comeback in the AEC industry.

Three Parametric Paradigms

As 3-D parametric software and tools are being rediscovered by architecture and engineering firms, they are beginning to change their design workflows. Contemporary design practices have developed at least three different narratives with regard to parametric design:

1. *Parametric formalism:* Parametric modeling and scripting has been used by a large number of digital avant-garde designers in intricate complex formal compositions [8]. Designers using this narrative use parametric techniques to substitute the manual designer in form-finding functions.

2. *Parametric BIM:* BIM has become one of the central themes in the computerization of architectural practice today. BIM software and processes allow architects and engineers to construct virtual models that accurately replicate building systems, materials, performance, and life-cycle processes. BIM narratives in practice have mostly concentrated in what the AEC industry calls 3-D, 4-D, 5-D, and 6-D BIM.

3-D BIM refers to collision detection models; 4-D BIM is used for construction sequence models; 5-D BIM models are associated with cost estimation; and 6-D BIM models are used for facilities management during the life span of the building. The merging of these parametric BIM models with embedded sensors procurement procedures, building simulation modeling, intelligent 3-D libraries, price engines, and bidding systems will move the narrative further. However, in spite of the exaggerated claims in the media that BIM is "revolutionizing" the AEC industry, BIM is still a labor-intensive procedure, and it is not a radically more intelligent method.

3. *Workflow parametric:* A third type of narrative is emerging inside design firms that are using parametric features to automate specific design workflows for projects such as façade design, environmental benchmarking, or structural optimization procedures. These groups are usually project-driven, part of special units inside the firms, and they work in aiding designers to explore generative and analytical computational processes in design.

Post-Parametric Era

Contemporary parametric metaphors found in scripting and BIM are only scratching the surface of a more profound transformation. Parametric allows for the coding of human reasoning. But parametric is still a manual, labor-intensive, and slow process. These systems are based on defining a large number of rules. However, anyone that has attempted to describe design processes with rule systems clearly knows that these systems get extremely complex after 50 to 100 variables are included. Parametric will not automate significantly design processes and will only slightly affect the economy of the whole AEC industry.

In Part II of this book we present a diverse array of cases of technologically progressive architectural and eEngineering firms that are at the forefront of this post-parametric era. The narratives of this post-parametric era are not singular or homogeneous, but on the contrary, they are very diverse and expanding every day. The major thread that brings together these firms are their questions about how they can further automate their own custom design workflows. These firms are moving beyond CAD/BIM/parametric modeling and into semiautonomous and algorithmically driven processes across different platforms to carry specific project tasks. Part II moves through a large array of case studies on algorithmically driven building simulation optimization, controlled façade shades, buildings, infrastructure projects, and urban design tasks.

Automating Architecture and Engineering via Machine Learning

In computer science, parametric is considered the most primitive stage of artificial intelligence (AI). As will be described in detail in Chapter 17, most of the major automation projects we see today in other industries are part of the second era of AI: the machine learning period. In this second era AI algorithms are no longer designed to perform particular tasks, but they are designed to learn without being explicitly programmed to do that task.

Machine learning algorithms are deployed to learn from data. They discover patterns and develop predictive behaviors or models to do particular jobs. In many industries these learning algorithms do tasks like the guiding of automated cars, the maneuvering of robots, or detecting patterns in data. AI algorithms allow apparatuses to perform tasks in real-time without being controlled by remote equipment or human. In Part II we show some extraordinary examples of how firms are moving into further automating their workflows as we move into post-parametric paradigms.

Automation Themes in Construction

From the late 1980s to 1999, large Japanese construction companies led the world in construction automation by building more than 550 systems [9]. These projects ranged from unmanned operations, robotics, avatar-operated equipment, and manufactured construction systems, to significantly automated construction processes. The Japanese experience has not percolated into the rest of the world.

Construction firms in the United States, Europe, and China have not introduced a noticeable number of automated systems as in Japan. Instead, they have preferred to focus on moving construction work into factory settings via prefabrication and modularization. In the past 5 years, a significant number of construction sites in the United States have become increasingly assembly sites in which elements such as heating, ventilation, and air conditioning (HVAC) systems, wall units, and even restroom components are prefabricated off-site, reducing safety, cost, waste, and the schedule of projects. In the United States, constructors' utilization of BIM technology also help further develop prefab imaginations. In one survey more than 70% of United States contractors contacted believed that BIM technology would allow them to increase prefabrication [10].

Part III presents several cases of automation from the construction perspective. One issue to note is that although all of these endeavors use prefabrication and/or digital manufacturing to some extent their main automation narratives are not directly linked to reducing labor on the job site. Sustainability, environmental concerns, design performance, material savings, shorter schedule, and better-quality products emerged as important motivators for prefabrication.

There are two major types of automation narratives in the construction process: manufactured prefabrication and custom fabrication. Manufactured prefabrication usually is led by major construction prefabricators who are using highly refined manufacturing and assembly systems to significantly reduce environmental impacts and improves the delivery process of construction. Cases of manufactured prefabrication are presented by the work of Broad Group in China in Chapter 13 and the Sekisui Heim Company in Japan in Chapter 14. Custom fabrication is typically led by architects and engineers interested in increasing design performance and quality. Cases of custom fabrication are presented in hospital construction by the architectural firm NBBJ in Chapter 15 and in the robotic fabrication of a pavilion at the University of Stuttgart in Chapter 16.

Automating Construction via the Future of Digital Manufacturing

The prefabrication and manufacturing automation narratives described in Part III are extraordinary but are by no means the ultimate image of automation in construction. On the contrary, they are just the preparation acts. Chapter 18 argues that digital manufacturing will ultimately challenge not only the way we process materials but also create completely new materials and eventually programmable matter—materials that can transform their physical properties via programmable control. The impacts of digital manufacturing will come in three different stages:

1. 3-D digitally manufacturing any forms;
2. 4-D digitally manufacturing completely new materials;
3. N-D manufacturing via programmable matter.

First, today an array of digitally controlled machines such as 3-D printers, CNC machines, robotic arms, and laser cutters is allowing us to manipulate any construction material with extreme accuracy. However, most of these impacts are at the level of manipulating materials at the human scale, but these changes do not affect significantly the performance of materials. Today we are entering into a 4-D digitally manufacturing era. In this second period we can use multimaterial printers and nanotechnology to manufacture completely new materials that cannot be found in nature. Further into the future a third epoch of N-D digital manufacturing will emerge when we are able to program materials to perform interactively based on evolving fields such as synthetic biology and evolutionary robotics apparatuses that are able to self-design and self-manufacture. We are far from entering into the mature stage of this third period but it is an important part of the narrative about how computer sciences might affect our analog world.

Conclusion

This chapter attempted to move forward a workable narrative about how the AEC industry is beginning to automate its workflows. There are two different narratives emerging in the forefront of automation today and these are very much related to the social units that led them. On the one hand we look at how a large number of architectural and engineering firms are transforming their practices by using parametric, BIM, and scripting tools that help them automate parts of their design and analytical routine work from design to fabrication. On the other hand we observe how large engineers/contractors have begun to transform their construction practices by moving gradually into prefabrication, modularization, and manufacturing.

Both narratives are incomplete. The design automation led by architects and engineers using parametric will not succeed in automating a significant number of workflows in the AEC industry. Instead, machine learning algorithms such as the ones used in many other industries will allow the design fields to automate their processes in a more effective way than parameter adjustments.

The current trend from engineers/contractors for prefabrication and modularization will potentially encounter the rise of 3-D multimaterials printers and synthetic biology processes. These methods can produce all types of new materials and biomaterials that can be designed at the micro- and nanometer level to respond to very particular conditions. This will lead to a completely new way of looking at digital manufacturing.

The advent of a more precise way of construction will eventually lead to a transformation of the designer and the traditional design process. Traditional design processes, either via hand-drawing or even with parametric CAD, are unable to plan with designing material performance at the macroscopic and microscopic levels. Machine learning design automation will have to play an increasingly important role in design synthesis for the construction elements that use multimaterials in the near future.

As was observed at the beginning of this chapter, automation implies important themes in saving labor, energy, and materials, as well as construction quality, and sustainability. The last factor will be an important factor throughout this book and the subject of the next chapter. The construction sector is in urgent need of modernizing and shifting toward sustainable construction practices as this has been identified by the United Nations (UN) as a key industry in the attempt to solve global warming [11].

References

[1] Andia, Alfredo. Managing Technological Change in Architectural Practice: The Role of Computers in the Culture of Design. Ph.D. Thesis, University of California, Berkeley, 1998.

[2] Noble, David F. *Forces of Production*. Transaction Publishers, 1984.

[3] Drucker, Peter. "The machine tools that are building America." *Iron Age,* August 30, 1976, p. 158.

[4] Gann, David M. "Construction as a manufacturing process? Similarities and differences between industrialized housing and car production in Japan."*Construction Management & Economics,* 14(5), 1996, 437–450.

[5] Crowley, Andrew. "Construction as a manufacturing process: Lessons from the automotive industry." *Computers & Structures*, 67(5), 1998, 389–400.

[6] Andia, Alfredo. "Reconstructing the effects of computers on practice and education during the past three decades." *Journal of Architectural Education*, 56(2), 2002, 7–13.

[7] Mitchell, William J. *The Logic of Architecture: Design, Computation, and Cognition*. MIT Press, 1990.

[8] Schumacher, Patrik. "Parametricism—A new global style for architecture and urban design." *AD Architectural Design*, 79(4), 2009.

[9] Obayashi, S. Construction Robot System Catalogue in Japan. Tokyo, Japan: Council for Construction Robot Research, Japan Robot Association, 1999.

[10] McGraw-Hill. Smartmarket Report. The Business Value of BIM: Getting Building Information Modeling to the Bottom Line. McGraw-Hill, 2009.

[11] United Nations Development Programme (UNDP) Report. Promoting Energy Efficiency in Buildings: Lessons Learned from International Experience, 2010.

Chapter 2

Green Automation: Design Optimization, Manufacturing, and Life-Cycle Sustainability

Alfredo Andia and Thomas Spiegelhalter

Introduction

Automating practice is a pathway of interoperable computation in the design and engineering workflow toward carbon-neutral architecture. In this chapter we argue that major international and national agreements that set new mandatory targets for achieving net-zero-energy buildings, to infrastructures, and cities by 2018–2030 are and will be a major driver of process automation with integrated project delivery in the AEC industry (Figure 2.1).

While there are a growing number of software applications and countless methods for writing custom applications and programs capable of leveraging the use of learning algorithms for many tasks within an automated design process, there is still a very limited understanding of how to integrate and adapt these capabilities into fully automated design-to-factory-file workflows. For instance, automation processes with feedback loop capabilities are natural partners to help designers improve the parameter inputs, predictions, optimize scheduling, identify patterns, and coordinate clashes and interferences. This also includes control and monitoring of inefficient energy and water systems in a building or even a city. In this example the most improved predictive systems are the most automated ones.

This chapter surveys the current generation of computational design optimization tools with interoperable whole-project analysis platforms, manufacturing, and building automation as they are currently used in the practice of engineering and architecture. However, the next generation of computational programming will begin to occur inside the automation domain and not in terms of software design.

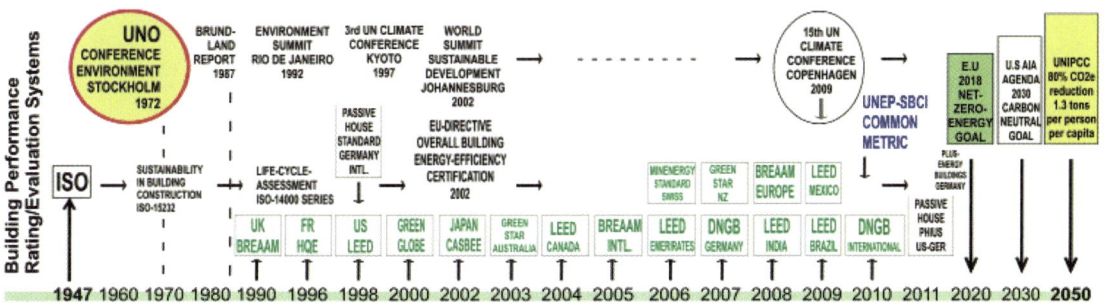

Figure 2.1 The evolutionary timeline of the worldwide implemented sustainability, building performance rating, and certification systems. (Source: Thomas Spiegelhalter [1].)

Toward Interoperable, Automated, Parametric/Algorithmic Carbon-Neutral Design Workflows

Worldwide, so-called net-zero fossil energy or carbon-neutral buildings and cities are still statistically pioneering concepts with some exceptional, mandatory, and national code and design protocol implementations in the European Union. In November 2009, the European Parliament and the European Commission agreed to recast the Energy Performance of Building's Directive (EPBD) from 2003 to make it mandatory that all new buildings in the European Union must use nearly net-zero fossil energy by 2018–2020 [1].

The targets for carbon neutrality can temporarily be accomplished through interoperable parametric-algorithmic design optimization processes to predict the future of the operational resource use of buildings. These design workflows also incorporate total life-cycle scenario tools for performance, material properties and resource use, and design-to-factory procedures. The intended interoperability for these building information model (BIM) platforms is the capability of autonomous, heterogeneous systems to work together as seamlessly as possible to exchange information in an efficient and usable way. The advantage is described that these 3-D-BIM design platforms links variables, dimensions, and materials to geometry in a way that when an input or simulation value changes, the 3-D/4-D/5-D model automatically updates all life-cycle scenarios and components simultaneously.

Some of those interoperable BIM platforms allow free plug-ins for several CAD tools (Graphisoft, ArchiCAD, Autodesk's Revit Architecture & MEP, Rhino, SketchUp, Grasshopper, Bentley, etc.). However, the major problems with these plug-ins are the inconsistencies in the noncompatible format exchange between different platform applications. Other limitations are the missing graphical human-computer interaction (HCI) user interface capabilities to allow easier and faster input and output of data with simple automated adjustments and improvements via learning algorithms.

For example, the current versions of Autodesk's BIM 360 and Green Building Studio (GBS) offer a Cloud-based service for architects that enables data exchange capabilities in gbXML format for automated building thermal geometry zoning, energy, water, carbon, and life-cycle analysis. The Cloud service engine imports any space type, usage, schedule, systems, components, and location. It automatically accesses over a million virtual real-time data-collecting weather stations worldwide. The analysis runs automatically through multiple parameters and algorithms of international, national, or local code compliance. Each of these engines generates predictive statistics and can compare baseline parameters with selected Energy Star, LEED, DNGB, UK-BREAAM, CASBEE or UNFCC-Carbon Emissions ratings for nearly all aspects of a building life-cycle during the design and planning process [2].

However, most of these Cloud services or BIM platforms for architectural design workflows depend—for example—on DOE-2, Energy Plus, or TRNSYS software algorithms and therefore inherit several of their problems and limitations. Some of the limitations are described and further developed in Chapter 10.

The next generation of system integrated platforms will be a type of inclusive automation, where computational programming and carbon-

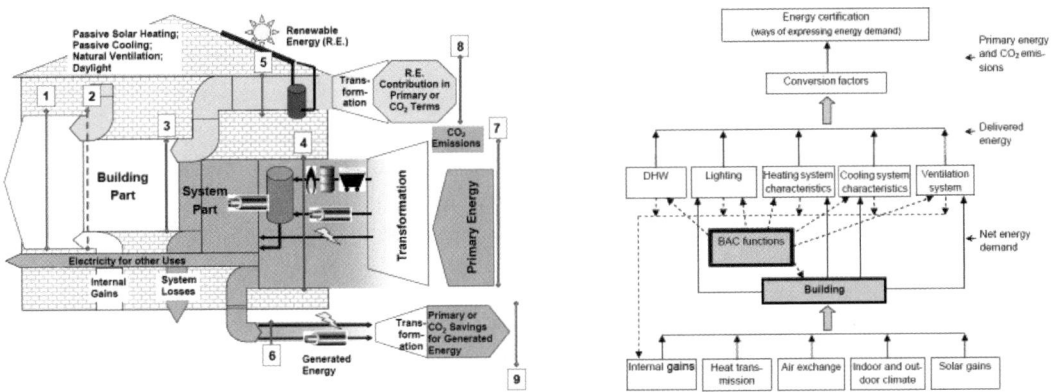

Figure 2.2 Diagram: Declaration on the general relationship between various European standards and the EPBD (Umbrella Document). (Source: Siemens AG.)

neutral manufacturing will be completely processed within the automation domain and not anymore in terms of computer systems. Designers and engineers will use flexible and easy graphical descriptions of the used system model and then there will be a more complex portion of software with integrated high-speed machine-learning and data analytics algorithms that automatically translate in real time new models into executable software. Another change that will dominate the future will be that the process of computation will be replaced by model-driven developments toward the use of conceptual models of applications rather than by concepts of computation.

In addition, the next generation of platforms will also include personal supercomputer systems and interoperable Cloud service worldwide. One example is the IBM super computer Watson, which got smaller and faster very quickly over a few years. According to BBC News "What started as a system the size of a bedroom is now the size of three stacked pizza boxes. It is also available via the cloud, meaning it can be accessed from anywhere. It can process 500 gigabytes of information—equivalent to a million books—every second"[3]. With such high-speed Cloud service supported supercomputers, sensor infrastructural polling in event-driven architecture simulation will eventually update or replace all the formentioned data exchange BIM platforms, which are currently only based on fixed or variable time step simulation concepts.

Chapter 12, titled "SIEMENS Digital (Self-Learning) Factories and Automation," outlines the latest parametric algorithmically driven multidimensional optimization tools in industrial design and in the automotive and transportation industries. The case studies feature SIEMENS PLM and Tecnomatix tools with integrated machine-learning data analytics algorithms and how they renew and optimize constantly the software models during design, manufacturing, assembly, and operation. The PLM capabilities offer open event architecture with multiple interface support, value stream mapping, and automatic analysis with constant optimizations of simulation and measured results to produce and deliver products and systems just-in-time (JIT) or just-in-sequence (JIS).

Total Green Building Automation System with Human-Computer-Interface Topologies

Today's building automation systems (BAS) are centralized, interlinked, and sensor driven human-computer-interface (HCI) networks of hardware and software. They monitor, control, and optimize in real time the environment in residential, commercial, industrial, and institutional facilities. While managing various building systems, the learning automation system ensures the operational performance (transportation, light, water, HVAC, energy generation, storage and distribution, etc.) of the facility as well as the comfort and safety of building occupants.

Historically, early generations of control systems were pneumatic or air-based and were generally restricted to controlling various aspects of HVAC systems in the 1960s to 1970s. These included controllers, sensors, actuators, valves, positioners, and regulators. The next generation shifted to analog electronic control devices with faster response and higher precision than pneumatics throughout the 1980s.

However, it was not until digital control or DDC devices appeared in the 1990s that a true automation system was possible. However, as there were no established standards for this digital communication, even though the automation system at the time was fully functional, it was not interoperable or capable of mixing products from various manufacturers. By the late 1990s and especially into the 2000s, movements around Honeywell, Siemens, or other major manufacturers were up to standardize open communication systems called BACnet, Ethernet, ARCNET, ModBus, LonWorks, KNX communication protocol that then became the industry open standards.

Today, most BAS operate with intelligent agents (IAs) and machine learning algorithms by identifying patterns for real-time optimization potential including time scheduling and trend logging and verification of building automation process. Intelligent agents in a BAS are sensors and effectors that interact with their environments. The systems topology of most BASs include the real-time generation of knowledge patterns and locations in multiple data scales that reiterate, change, and optimize automatically new building energy, resource, security, circulation peak load, and user comfort management processes.

For example, Siemens uses wireless, automated, self-learning two-position algorithm sensor infrastructures that constantly control and fine-tune building spaces and zoning conditioning demand. Today, fully integrated multidimensional trend data processing allows effortless event-driven polling and analysis of real-time (online) data and (offline) historical data in compliance with multiple standards. Any energy/water/resource use and cost reports including CO_2 or net-zero-energy building (Net-ZEB) benchmarking values can be assembled and polled in real time at any time during the operation of buildings.

The future of green building automation will be Cloud-computing-controlled buildings. Cloud-controlled buildings provide the flexibility to expand wireless infrastructures with sensor-collected trend data and self-programming data analytics algorithms. The Cloud will be where the applications run and where the data is analyzed and acted upon as it arrives. Digital data is changing; we are moving into a world with an ever growing number of data sources. As the amount of the data and the requirement for algorithms that act on the fly increase, a green BAS cloud will be able to automatically do real-time stream analytics of different variables in seconds and expand itself to accommodate the operation and peak load control needs on any scale from buildings to cities.

In Chapter 8 the Q1 Thyssen-Krupp headquarter case in Germany describes how a real-time SIEMENS total green building automation system (BAS) performs with intelligent control feedback loops and learning algorithms for constantly optimized building performance, security, and user comfort operation. This system also includes a wireless environmental management system to ensure trend analysis and optimizations toward yearly mandatory net-zero-energy certifications. In Chapter 3, we describe Broad Group's 30-story hotel building automation system that is an essential part of their prefab sustainable building strategy that includes a high insulation approach. Their BAS monitors all the sensors and controllers in every room of the building with the overall building systems to maintain a critical balance of air circulation and air purification strategy with a low-energy approach that according to the designers uses 20% of the total energy consumption (per primary energy) of comparable buildings.

Also in Chapter 5 the Al-Bahar Towers algorithmic principles illustrates the control software and building management systems with a human machine interface software that was developed by the Al Bahar tower engineers using the Siemens and supervising control and data organization (SCADA) product. For the parametric design 15 different software packages were used by various parties to develop and deliver their scopes to feed data into the CNC machines for fabrication. Topographic survey machines on site were then utilized for installation and later for the building automation for constant operation performance benchmarking.

Automation in Green Building Manufacturing

Today and in the future, automated green building manufacturing will go naturally together with faster and more flexible customization, and corporate sustainability strategies to reduce manufacturers' carbon footprints and energy costs for revenue growth with return on capital employed (ROCE). In this context many national environmental protection agencies around the world are already mandating greenhouse gas reduction and mitigation reporting rules requiring manufacturers to file annual emissions reports to bring them into compliance.

For example, a typical car manufacturing facility "with a daily output of 1,000 vehicles consumes several hundred thousand megawatt-hours of electricity per year—as much as a medium-sized city. The electric motors used to drive conveyor systems, robots, and other machinery use two-thirds of this power, and optimized control systems can reduce their consumption by as much as 70 percent" [4]. Many studies are on the way to make industrial robots, conveyors, and transportation lines more energy efficient by simply automating the software that controls and self-optimizes their movement patterns which can save up to 30 to 40 percent in energy and CO_2 mitigation costs. There are also increasing design/built examples of green manufacturers in the AEC industry where these facilities are already completely operating as net-zero-energy or carbon neutral entities.

Another example is that flexible automation with self-learning robots in mass customization will also usher in a new era of green choice and flexibility for manufacturers and clients in the AEC industry. Sustainable traditions from the craftsman era that were either lost or underscored during the era of mass production can now be individually integrated in green manufacturing and 3-D printing settings.

Over the next couple of decades, we will see major enhancements in automated scenario network planning and in high-speed cloud computing that will further improve resource innovations and flexibility. Fully automated production control and optimization will boost factory productivity. With fewer inputs to make more outputs, managers and production workers will naturally still be in charge, but they will be controlling automated software and processes rather than the self-learning machinery, robots, and sensor-driven intelligent agents. Increasingly, 3-D printing technology will create complex building materials, components, and systems in multiple programmable scales. Even further advances in multidimensional printing technology scales are enabling mass customization at increasingly granular levels. Most of these game-changing processes are described in further detail in Chapter 18.

In general, what we now have are firms that are truly committed to more sustainable approaches, such as Broad Group (Chapter 13) and Sekisui Heim (Chapter 14) radically transform their manufacturing processes. In doing so they are forced to rethink the most basic principles of traditional construction by doing more with less materials, less waste, fewer trips of construction vehicles to the job site, and all this for a cheaper price and a much lower carbon footprint.

Conclusion

In this chapter, we presented a brief overview of green automation, which has been applied for design optimization, manufacturing and life-cycle sustainability. Of course, the related works presented here are neither complete nor exhaustive but only a sample that demonstrates the value of green automation and self-organizing systems. In summary, software architects have migrated from the old error-prone paradigm of programming to the "new world of system integrated and model-driven development—that is, the use of conceptual models of applications rather than computing concepts" [5]. In the future, computational programming will happen in terms of the automation domain and not in terms of computer systems. The next generation is a type of green automation, where designers and engineers deal with graphical descriptions of system and complex cloud software with machine learning algorithms that automatically repeatedly translate new models into optimized executable software. We are on the verge of a paradigm shift, where "communities of machines will organize themselves, supply chains will automatically coordinate with one another, and unfinished products will send the data needed for their processing to the machines that will turn them into merchandise" [6].

This new era of green automated virtual-to-real manufacturing will reorder the global AEC business for decades. The AEC industry that capitalizes on these changes across its entire development, production, and building post-occupancy benchmarking processes will set a tone in which others will be challenged to follow in order to remain competitive.

References

[1] Thomas Spiegelhalter. "Achieving the net-zero-energy buildings 2020 and 2030 targets with the support of parametric 3-D/4-D BIM design tools."*Journal of Green Building*. Spring 2012, Vol. 7, (2), pp. 74-86.

[2] Energy Analysis Software—Green Building Studio—Autodesk, http://www.autodesk.com/products/green-building-studio/overview, retrieved on 27 February 2014.

[3] IBM seeks app developers to harness Watson, BBC News, http://www.bbc.com/news/technology-26366888, retrieved on 27 February 2014.

[4] SIEMENS Maximizing Efficiency—Efficient Car Production http://www.siemens.com/innovation/apps/pof_microsite/_pof-spring-2013/_html_en/energy-management.html, retrieved on 27 February 2014.

[5] Lothar Borrmann. "Making sense of complexity," *Siemens—Picture of the Future*, Fall 2013, p. 14.

[6] Katrin Nikolaus. "Building the nuts and bolts of self-organizing factories." *Siemens—Pictures of the Future,* Spring 2013, p.19.

Part II

Post-Parametric Workflows in Architectural and Engineering Offices

Alfredo Andia and Thomas Spiegelhalter

This chapter provides extraordinary examples of how architectural and engineering firms are semiautomating some of their important design workflows. We are in a transitional period in a post-parametric time. Parametric and CAD/BIM platforms allowed architects and engineers to organize and analyze form in digital environments. But as different technologies have infiltrated significant aspects of practice, today's designers are asking higher-level questions: How can the design workflows be simplified and automated? How can automation procedures assist in increasing the number of candidate design solutions in shorter and more complex design cycles?

However, automation doesn't appear suddenly and it is an evolving computerization theme that comes in multiple platforms and with a growing number of narratives. Designers are observing the algorithmic automation themes emerging in other professions. Today there are intelligent algorithms in our phones, cars, traffic management, electricity distribution, robotic apparatuses, personal supercomputing, big data analysis, and many other spheres. Design professionals are beginning to ask to what extent will these automated algorithms continue to infiltrate into design domains.

Social units in the AEC industry are organizing both collectively and individually to answer these questions and to further expand their ideas of practice. The narratives of this post-parametric era are not singular or homogeneous, but on the contrary, they are very diverse and expanding every day. At present, a highly technological design firm could be working with 100 or more applications at the same time. The emergence of scripting and algorithms in design processes is making the computerization stories of practice even more diverse. This section attempts to reflect a significant range of the discourses that are present in this post-parametric period in the AEC industry.

In Chapter 3, Keith Besserud from Skidmore, Owings, and Merrill (SOM) in Chicago, discusses several design/built optimization processes within the practice of his firm. At SOM there are search algorithms that work like an automated sculptor that removes piece by piece the material that is not needed in a 3-D model in the process of designing structural trusses for skyscrapers. Genetic algorithms are used in performative searches for structural and environmental control solutions and metrics. SOM is also using the assistance of algorithms to tackle large-scale urban systems-based projects, such as a 600-acre development in Chicago's South Side where they are testing a virtual urban design environment named LakeSIM.

Chapter 4, by Christian Derix, founder and former director of the Computational Design Research group (CDR) at Aedas, describes the foundations of 10 years of research and design/built projects at one of the research and development arms of one of the largest architectural firm in the world. CDR has focused on developing algorithmic and heuristic design methodologies that provide architects and stakeholders with new representations of space. CDR is interested in the computability of design. They have developed a large number of highly innovative in-house heuristic digital models that aid designers in their design search. For example in one approach they used self-organization-based agent models with attract-repel algorithms in which a user can interactively generate space planning and quick massing studies. Other methods include urban spatial planning, access design, and occupancy and behavioral mapping. The chapter emphasizes that architecture is meant to provide experiences by using spaces and observes that digital design procedures should be able to help generate, visualize, and evaluate the heuristics of places and users.

Chapter 5, by Abdulmajid Karanouh, concentrates in detail on the generation of 1.049 kinematic folding daylight redirecting and shading screens that interactively react to the sun path for two large towers in Abu Dhabi. The project was initially a competition proposal developed with the CDR unit at Aedas described in Chapter 4 and it shows how the computation themes developed by an R&D social unit began to percolate in design projects. The design, development, and manufacturing of this project demonstrates the synthesis of Islamic and regional architecture plus sustainable technology with the inspiration from nature to develop an algorithmically driven design-to-project delivery strategy.

A series of bespoke learning algorithms were developed following underlying mathematical principles inspired by the universal order of orbital motion to realize a microclimatic and automated adaptive enclosure system for the office tower. It is notable that at no time was the parametric-algorithmic scripting and design-to-fabrication process limited to a single CAD/BIM/parametric platform, which allowed the experimental use of over 15 different software packages.

In Chapter 6, Thomas Spiegelhalter presents the internationally awarded 1 Bligh Street, Sydney, Green Star rated high-rise project resulting form the collaboration of Ingenhoven Architects, DS-Plan from Germany and Architectus, Arup, Enstruct, and the builders Cundall and Grocon in Australia, with the typical challenges and problems of the firms in the multidisciplinary CAD to BIM collaboration. The firms used different methodologies and approaches, producing different input formats for the 3-D, 4-D, and 5-D BIM platforms with altered output levels of details and system scales for repetition in parametric modeling and automation in the design-to-fabrication processes. Most automated processes were executed by the structural engineers and contractors by rationalizing a series of circular arcs of the building systems which then could be mirrored and automatically repeated in the design-to-fabrication processes.

In Chapter 7, Lucio Blandini, Albert Schuster, and Thomas Spiegelhalter illustrate how a large-scale infrastructure project is designed, coded, and scripted through a highly automated workflow process of nonlinear analysis and structural behavior optimization methods. Scripting was hereby a very helpful method for the automated modeling and optimization of all the workflow scenarios between the different professionals involved. Besides the structural optimization, the project was also algorithmically modeled to discover the most efficient low-energy scenarios and assembly strategies. Compared to an average railway station structure with the same spans, this team was able to reduce the overall structure to one-hundredth of span, resulting in the use of much less material. The new zero-energy railway station is discussed as a prototype of a new generation of railway typologies that will provide passenger comfort with passive strategies on the highest level.

The net-zero-energy Q1 Thyssen-Krupp Headquarter, Essen, Germany, in Chapter 8, analyzed by Thomas Spiegelhalter, is an example where first a linear approach, originating in sketches, 2-D plans, and then proceeding into nonlinear 3-D digital master model workflows and digital mock-ups with file-to-production. The next shift occurred when the complexity of the project demanded real-time 3-D simulations with VRfx compatible formats in OpenGL Performer software to share, reiterate, synchronize, and visualize quickly changes and updates in collaboration with the direct input of Thyssen-Krupp AG (client) and their specialized contractors. More than 300 companies and 50 involved planning firms were also coordinated through an additional information life-cycle management (ILM) data platform to cover all processes of planning and construction throughout the life-cycle and financial management of the real state. Thomas Spiegelhalter observes that the production, transportation, and assembly of the highly adaptive building enclosure systems were executed through automated bar-code-controlled just-in-time supply chain processes. The project also includes, besides the collaborative, real-time OpenGL Performer analysis and scheduling, a real-time total green building automation system (DESIGO) based on intelligent control feedback loops with self-learning algorithms for constant optimized building operation and benchmarking.

Chapter 9 by Clayton Binkley, Mathew Vola, and Paul Jeffries from Arup Firm Seattle, Brisbane, and London respectively present the design computation processes at this engineering firm. The vast majority of Arup's engineering work today is highly intertwined with computers. As the authors state, "much of our computation work is simply automating customized design processes such as: design checks, model interrogation, data harvesting and processing and automated documentation or visualization tools." The chapter describes in detail how they use, adapt, and develop custom software, algorithms, and scripted workflows to automate significant parts of the design-to-fabrication processes. They present in detail their automated workflows in two major projects in China and Japan in which they combine a large number of software and algorithms further blended with engineering design intuition in order to realize highly complex physical objects.

In Chapter 10, Lars Junghans elaborates in detail a large number of different building optimization algorithms and how the building sector could move into an automated building optimization paradigm. He compares the current and future use of enumerative search methods where all parameters are combined with each other to use automated, multiobjective building optimization algorithms coupled with software platforms to find optimal scenario solutions. His observations include critical insights about the speed of calculation time and questions whether optimization algorithms can be used by architects and planners without expert knowledge in optimization theory and computer science. The article concludes with the case study of a six-story office building project constructed in Austria in 2013 in which the author was in charge of the comprehensive design of the energy concept. The building is unique because it has no active heating, cooling, or ventilation system in a very cold climate. All the energy flow and space conditioning systems are controlled by a sophisticated software building automation system (BAS) with self-learning algorithms.

Chapter 11 by Lucio Blandini and Werner Sobek showcases a parametric and semiautomated engineering, manufacturing, and assembly workflow for the construction of the Enzo Ferrari Museum in Modena, Italy. The museum was designed by Jan Kaplicky of Future Systems from London shortly before his death. The 3-D modeling of all the systems and components needed to be precisely coded and scripted through a highly automated workflow process of nonlinear analysis and structural behavior optimization methods. All the elements were designed and manufactured specifically for the Ferrari Museum's language, with the aim of reducing the material used to a minimum and to match the specific dynamic, free-form, high-performance automotive and architectural vocabulary.

Chapter 3

Engaging with Complexity: Computational Algorithms in Architecture and Urban Design

Keith Besserud, SOM

Introduction

In the course of designing buildings and cities, architects and urban designers quickly confront inherent complexities of at least two very different natures, one relating to the design process itself, and the other relating to the subjects of these design processes (the buildings and cities).

First, there is the tacit understanding among designers that the field of all possible design solutions (the solution space) for a given design problem is far greater than the design team will ever have the opportunity to fully explore. Given the time limits of a typical design cycle, the design team will only have the opportunity to conceive (let alone rigorously interrogate) a very small fraction of all the possible approaches to designing the building or city. Equally unacknowledged is the fact that because this sampling is so small, the team can claim little substantive knowledge of how good the design actually is (as defined by whatever metric you choose) compared to those undiscovered designs within the solution space that are actually the best.

This is the reason we typically enlist large numbers of (usually young) designers in the earliest stages of the conceptual design phase so that we can efficiently canvas as much of the solution space as possible, in hopes of discovering something that an experienced designer can intuitively recognize as promising.

The second type of complexity that the design community has been forced to marginalize is systems-based complexity. Although the number of systems that make up buildings and cities is relatively finite (though still extremely large), the number of interconnections that feed from each system into the others, and which may vary over time in magnitude, quickly accumulates exponentially into a staggering number of domino events that are also impossible to keep track of within the time constraints of the design process. In the face of this impossibility, designers have historically been forced to routinely simplify the problem, to disconnect the interdependencies between systems models, to use rules of thumb, and/or to make various assumptions as they iterate through the design space.

Recently, however, computers and parametric modeling platforms are allowing designers to manage and engage with some of these forms of complexity in ways that have never been possible before. In particular, computationally driven strategies for conducting searches of large design spaces and for capturing complex systemic relationships are beginning to emerge within the design professions. Not only do these types of tools allow for better management of the complexities of our design problems, they can even be leveraged to drive those design processes.

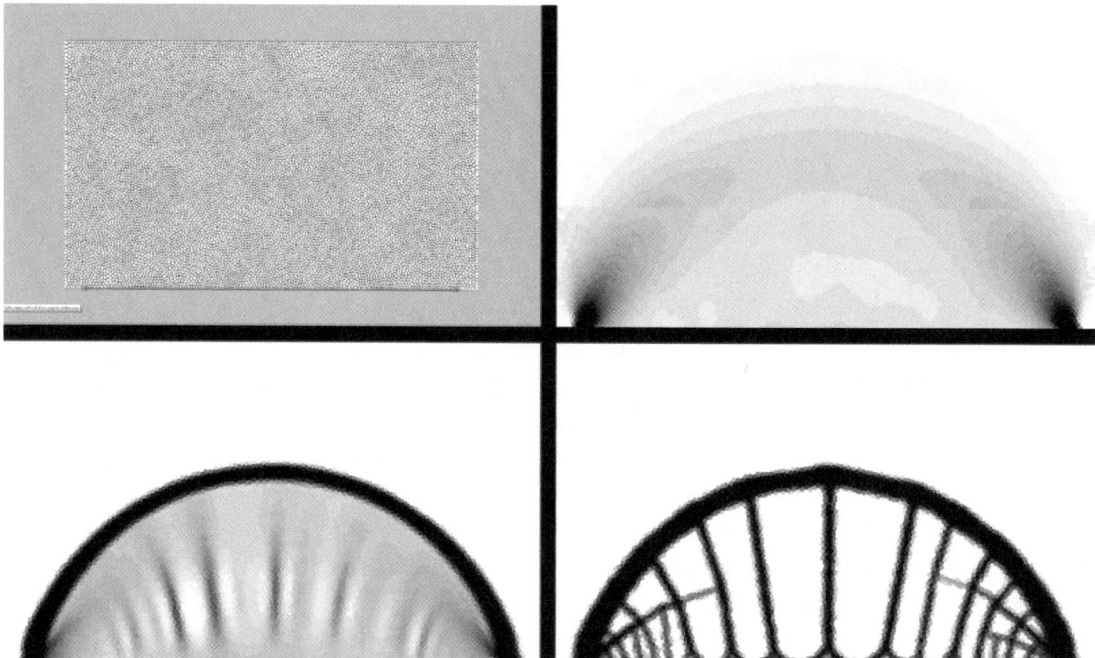

Figure 3.1 Optimal truss research: This series of images reveals the optimal structural form for a truss, given a specific set of loading, restraint, and meshing criteria, generated using a force-density algorithm.

Search Algorithms

One example of this new ability to engage with complexity can be illustrated in the use of search algorithms. Typically, this term brings to mind search engines for the Web that can generate lists of Web sites that will presumably contain useful information for the user based on the search criteria the user provides. The alternative is for the user to try to think of all the Web sites they know that might have the information they are looking for and then manually navigate to each site to see if it has relevant information. Obviously this would be very inefficient and also not very effective.

Search algorithms can be used to overcome similar limitations in design processes. Typically this process is known as optimization, but it is fundamentally a search process. Given that the solution space contains far more designs than can be sampled, these algorithms are designed to automate a process of sifting through very large solution spaces to find very good solutions. The effectiveness of these algorithms is highly dependent on the designer's ability to define "very good" in terms the computer can understand and/or calculate; as a result they currently tend to be leveraged more frequently in situations where the designer is interested in certain types of performative qualities that are more easily measured, such as structural, sustainability, or financial.

An ongoing research initiative in our structural engineering studio is exploring the concept of efficient truss topologies. One strategy being leveraged in this exploration is an optimization method called the force-density method. In this search technique, the solution space is a somewhat arbitrarily defined block of material with explicitly defined supports and applied loads. The search algorithm works like a sculptor, iteratively removing pieces of the block of material that are doing the least amount of work in the transfer of the forces from the loading points to the

Figure 3.2   Left: Results of force-density analysis showing "high-waisted" cross bracing topology. Center: Image of Sydney tower. Right: Image of John Hancock Center.

supports. The result of this iterative subtraction process is a truss form that corresponds to the most efficient use of the available material; in other words, a structurally optimal truss. If the loading or support conditions are changed, the resulting form will be changed as well.

This same technique influenced the design of a project for a tower in Sydney, Australia, that expressed a structural concept called a braced frame. Perhaps SOM's most iconic use of the braced frame concept was on the Hancock Center in Chicago in the late 1960s. At that time it was assumed that the most efficient form for a cross brace was two straight members that crossed at midspan. In fact, through the use of this same search technique it was revealed to designers that a more efficient form for the cross brace was one in which the intersection point was elevated some distance above the midspan intersection point. This high-waisted brace was used in Sydney, effectively representing an evolutional advance over the thinking from the Hancock project.

Genetic Algorithms

Another search technique being used on design problems is the genetic algorithm (GA). With a GA, the designer must first characterize the design as a series of numeric variables that correlate to the geometric properties of the design, essentially its genome. These values are free to

Figure 3.3   Use of genetic algorithm on Ali Al-Sabah Military Academy design process. Left: Overall birdseye view of campus. Right: Image of window design generated with the genetic algorithm.

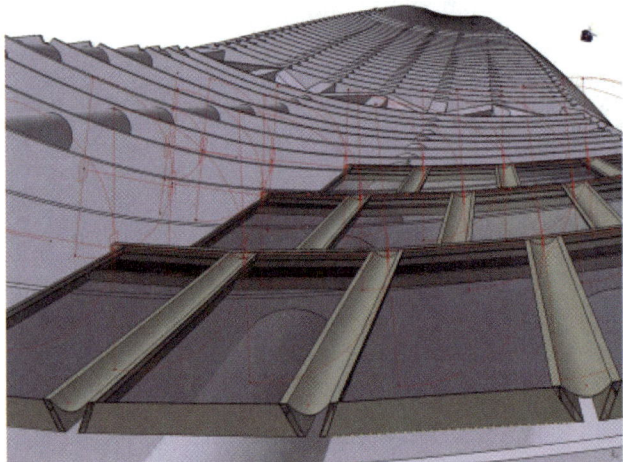

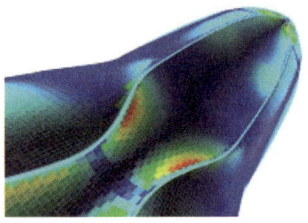

Figure 3.4   Integrated modeling of the building enclosure systems. Upper left: Parametric model of a tower in China, showing curtainwall, structural slab, and perimeter column components. Right: Artistic rendering of tower. Lower left: Results of curtainwall analysis that colors each panel according to the amount of bending that is required to maintain the façade curvature of the design.

change during the course of the algorithm's execution as it searches for the combinations of values that lead to the creation of the best solutions. Again, the designer must also describe "best" in terms the computer can calculate and measure.

Whereas force-density algorithms are designed specifically for structural explorations, GAs can be used for any kind of performative search for which there is the ability to computationally measure that performance criteria. Although we have used GAs for structural explorations, we have also used them for searches oriented toward solar radiation and daylighting metrics. One example is the Ali Al-Sabah Military Academy in Kuwait. Here we used the GA to help us identify a range of well-performing window designs for each of the primary wall orientations. Well-performing in this case meant that the window geometries did a very good job of minimizing direct solar gains while also maximizing the amount of indirect natural lighting.

Systems Modeling

The second domain of complexity we are engaging with more directly is that of integrated systems, at both the architectural and urban scales. Architecturally, one particular area of growing interest to both architects and engineers is the zone of the exterior skin assembly, especially as the shapes of the building designs are becoming more geometrically complex (no longer just simple rectangular extrusions). At the urban scale, as sustainability issues (especially relating to energy and water) become more pressing, our urban designers are taking a much greater interest in the design of the infrastructure systems of the cities. Both of these cases—the

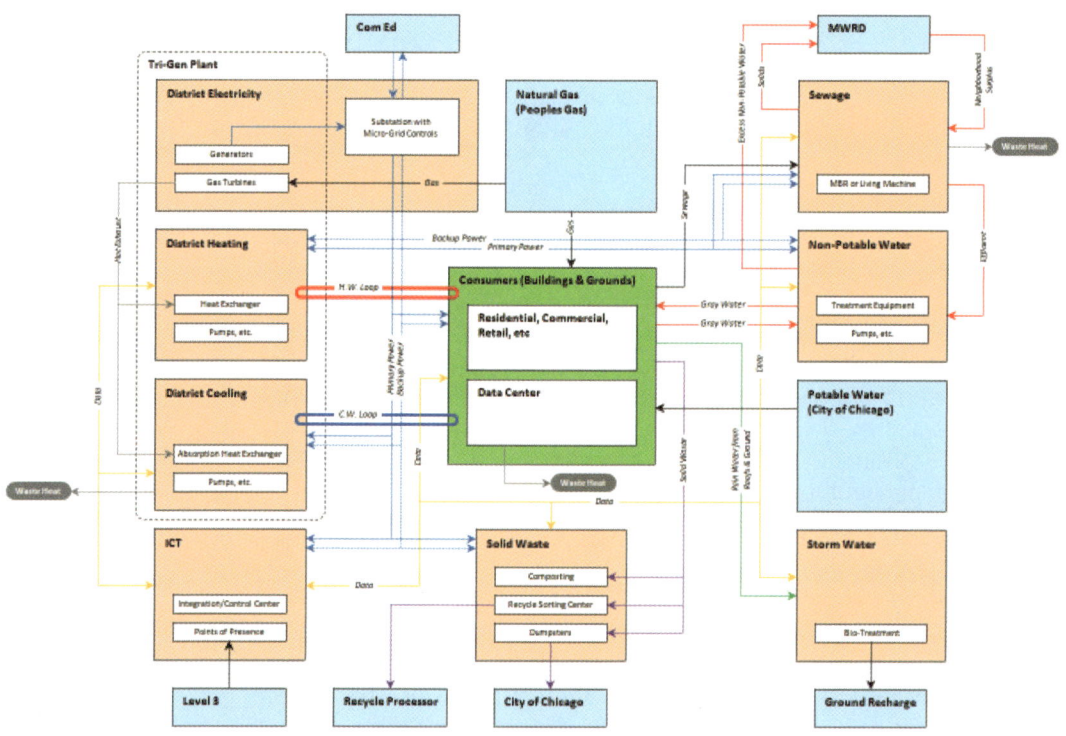

Figure 3.5 Modeling of urban infrastructure systems. Top: Schematic diagram for the proposed LakeSIM modeling software, showing the flows of resources and interdependencies of the various utility systems for a proposed urban design project. Bottom: Artistic rendering of the urban design project called Lakeside, Chicago.

exterior wall assembly and the utilities infrastructure—represent examples of systems complexity that our design teams are now engaging with greater rigor, thanks mainly to advances in computational capabilities.

With built-in palettes of NURBS-based shape generation tools, geometry modeling programs like Rhino, Grasshopper, and Digital Project have made it relatively easy for designers to create building shapes that are highly complex and curvilinear. These complex curvilinear geometries must then be rationalized into curtain wall cladding systems, generally consisting of quadrilateral panels of glass and aluminum framing systems. Additionally, the structural systems (including the floor slabs, perimeter columns, and substructural components) that hold up the curtain wall system must also be rationalized to reflect the surface geometry of the building.

In fact, the skin and the structure for a high-rise building are tightly integrated systems, even though the design teams responsible for each are highly distinct and are usually responsible for their own models. When the building's exterior surfaces are plumb and planar, it is not so difficult to coordinate and integrate the designs and models of the two systems; however when the shape of the building becomes much more complex, coordination and integration become much more challenging. Perhaps the biggest driver of this challenge is the fact that the shaping of the building, as well as the positioning and sizing of the structural components, may remain in independent flux for much of the design process.

Over the course of time and multiple projects we have come to develop a number of processes and standards to better ensure the coordination of the architectural model of the curtain wall system with the structural model of the slab edges and perimeter columns. The keystone of the system is the master geometry file that describes the idealized exterior surface of the building, usually in Digital Project; all the rest of the geometric data is derived either directly or indirectly from this surface geometry. Slab edges are derived by first cutting planes through the master surface at the elevation of each slab and then generating an offset of that curve at some distance inward. The columns are positioned relative to the slab edges, although the precise location is not known until they are sized because they are positioned relative to their surfaces, not their centerlines. The most important numerical parameters are hosted in a central spreadsheet or database, which the architects and engineers both reference in order to ensure identical geometry for both models. Ultimately the 3-D structural models are reconstructed in the architectural model so the geometry is really a hybrid of both architectural and structural inputs.

Complexity is of course abundant at the urban scale as well, where we find that sustainability motives are driving urban designers to take a much greater interest in the design of utility infrastructure systems. The two metrics of greatest interest are usually the carbon footprint and the life-cycle costs (LCCs). For the most part, however, these metrics have only been looked at within the isolated contexts of each individual system, meaning that we have been able to study the carbon footprint and LCC of the electricity infrastructure, but could not very easily wrap our arms around the consequences, for example, of different choices about power generation technologies on the other utilities, like water, natural gas, and waste heat. The various options for the componentry of each system—combined with the different input/output specs for each component choice—lead to a dizzying array of possibilities that can quickly overwhelm the design team.

LakeSIM: Integrated Infrastructure Modeling Platform

In order to design these types of integrated systems of systems, designers need better tools for managing the information they are working with. This is the motivation behind a joint initiative between SOM and computer scientists at Argonne National Lab to develop an integrated infrastructure modeling platform called LakeSIM. The goal for this project is to create a 3-D geospatial design environment in which the designer can parametrically create and manipulate different urban design scenarios, including street grids, building masses, land use designations, and utility infrastructure systems, and get near real-time feedback for whatever metrics are tracked as modifications are made to the design.

In theory, for example, the designer could change the density or the land use designation of a block within the LakeSIM environment and immediately see how this change affects demand on the electricity system, which could in turn affect the size of the power generation plant, which could in turn affect the natural gas demand that runs the turbines as well as the amount of waste heat coming off the generation plant that is available to supply the district heating and cooling needs of the homes in the neighborhood. All the domino effects that are tripped by all the systemic interdependencies would be calculated automatically. It is easy to see how trying to manage all this dynamic information manually would quickly become impossible, but with the help of a model like LakeSIM it is possible.

Conclusion

Buildings and cities are highly complex propositions, both in terms of the deeply integrated systems of systems they encompass in their operations as well as the myriad of parameters, constraints, and goals that must be reconciled in the process of designing them. Historically, our ability to represent this complexity, to manage it, and to actually allow it to inform the design process has been greatly constrained by the time allocated to the process, the tools available, and the limited carrying capacity of the bare human intellect.

Time allocations and our mental carrying capacities have probably not changed much, but the tools and the models certainly have. Over the last couple of decades, and especially in the last few years, we are seeing that computers and software tools such as search algorithms and integrated systems models can prove extremely useful in helping designers get a much better grasp of certain types of complexity and to extract crucial forms of information from that complexity.

Chapter 4

Space Planning with Synthetic User Experience

Christian Derix

Introduction

Traditionally, media of representation in architecture such as perspective construction have served foremost to communicate conditions and properties of space that the user is employing to imagine being there in that space/place and to evaluate design decisions. The medium of computation has largely deviated from space as the core subject of architecture. This is due to a series of opportunities that computation provides at first sight, three of which are (1) quantification with large amounts of structured data, (2) fast processing of explicit procedures, (3) leading to optimized performances. These opportunities are incongruent with the polyvalent and implicit nature of architectural and urban space as Rittel pointed out by calling spatial design tasks "wicked problems" [1], as no explicit linear solution path to a complex spatial state is possible. Hard delimitation that renders spaces visible and quantifiable like envelope or physical infrastructures receives the most attention, both in academic research and professional development. Here parametric representation of spatial boundaries and control via BIM for example, has created obvious acceleration and precision for describing explicit performances. This has led to an objectification of architecture on par with industrial design, hence the incessant comparison with the automotive industry, reflecting a modernist mentality.

Research into user-centric representations of space exist, particularly in the United Kingdom: syntactical analysis of spatial configurations that give rise to affordances of occupation by Philip Steadman and Bill Hillier and epistemological models of generative design by Paul Coates and John Frazer. Syntactical analysis investigates the user as occupant through correlations between spatial configurations and the potential for occupation, while epistemic generative models focus on the user as designer or the emergence of spatial organizations. Both aim at exploring representations of an agency of the user in space.

So far those fields have remained largely isolated from each other and the profession. They did not appear to add value to the architectural industry, certainly not in standards of efficiency or cost. But during recent financial difficulties and densification efforts, the value of space as both urban and property as commodity has increased for which methods of adaptability and value-addition need to be formulated. Formal embellishment by way of parametric modeling is slowly giving way to spatial performances, simulated via configurational, operational, and occupancy models.

During 10 years from 2004–2014, the Computational Design Research (CDR) group of AedasIR&D focused on developing design methodology that would provide architects and stakeholders with new representations of space. Both types of models—syntactic spatial analysis and epistemic generative design—have always run in parallel and recently also synthetically, being the first models that integrate the two approaches. Below, some background and developments of the work of CDR are presented.

Space as Heuristic Organization

Space represents a mapping of actions, be they of the designer or occupant. Computation as a notational system generating diagrams of conditions according to Stan Allen "goes beyond the visual to engage the invisible aspect of architecture," beyond climatic and sensory qualities, "perhaps more significantly – program, event and social space, ... The use of notation marks a shift from demarcated object to extended field" [2]. The extended field represents the implicit space of relations between and including physical boundaries (i.e., the space of actions). It's a metageometric space where users and geometric representations interface, lending space its affordances and thus function.

This interdependence is encoded by at least two processes: the generative process and the enactive process. The generative process of "designing space" encodes the heuristics of the architect, while the enactive process decodes the heuristics of behaviors by occupants. Heuristics are learned methods of correlating design actions with desired spatial conditions (rules-of-thumb for certain design situations), which mostly rely on transmitted design thinking and best practice (Figure 4.1). Occupant behaviors are learned from cognitive responses to spatial situations (experiences) and are triggered by cues from spatial conditions. Designers use heuristics to generate spatial configurations while the occupant cognitively interprets spatial configurations to take actions, without the two being causally related.

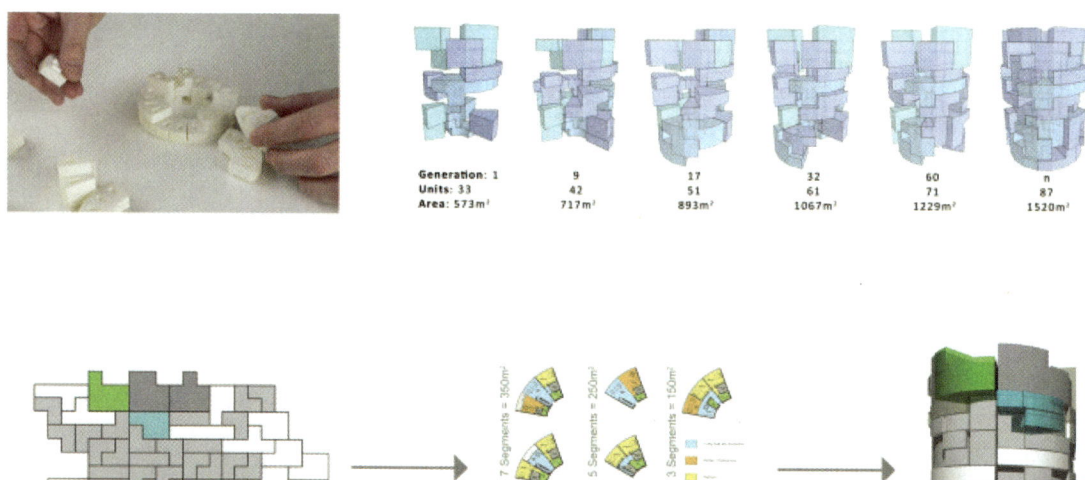

Figure 4.1 Khalifa-bin-Zayed competition, Abu Dhabi, UAE (2009): CDR developed a series of design applications that enable designers to pack 3-D polyomino apartments that would be practically impossible to solve by hand for their spatial complexity. (Copyright Aedas architects.)

Models of Designer as User

A designer learns how to organize spaces according to his or her cultural setting, and rules of composition are often distilled in design guides. For example, the main urban design guide in the United Kingdom is the Commission for Architecture and Built Environment's ByDesign [3] where design objectives and spatial aspects are discursively described. Such guides and standards provide implicit measures and methods for good practice, aiming to inform the heuristics of designers. Most aspects of spatial form are reminiscent of perceptive qualities such as those described by Kevin Lynch in his book *The Image of the City* [4]. His measures refer to qualitative distinctions that derive from observations and experiences in the field.

In computing, models can be developed that evoke cognitive identification by the designer with the visualized computing process. In other words, if a model can be observed to display apparently intentional behaviors akin to learned design heuristics and objectives, a tacit agreement between model and designer is created. Frei Otto's catenary models, albeit analog material computation, displayed such behaviors [5]. Otto and his designers could observe behaviors and spatial states that were epistemologically autonomous from the designers yet aligned with their intentions. Seymour Papert called such models that encode a user's abstract knowledge ego-syntonic [6]. Computationally such models are possible by creating lean representations of processing behaviors by reducing the number of parameters and procedures for computation. This reduction also enables the creation of legible diagrammatic visualizations [7].

In 2009, CDR designed a heuristic design model for the Abu Dhabi Education Council competition, which would aid the architects to explore the design space of the building capacity and mass. Like Otto's catenary model, the simulation always represented a good equilibrium within the solution space. For the approximation of the building mass, a self-organizing agent-based model was developed based on the attract-repel

Figure 4.2 Abu Dhabi Education Council competition, Abu Dhabi, UAE (2009): The formal input to computational systems should be as diagrammatic as possible (left); a manually processed outcome from the computational system. (Copyright Aedas architects.)

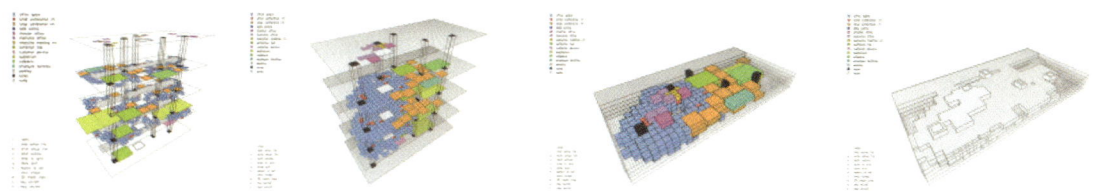

Figure 4.3 Abu Dhabi Education Council competition, Abu Dhabi, UAE (2009), from left to right: Initial random placement of accommodation units; interactive semiautomatic ordered layout; block model visualization; second skin for final envelope extends. (Copyright Aedas architects.)

algorithm that simulated the search for relationship between room units, specified in the adjacency matrix of the accommodation schedule. Apart from adjacencies, room sizes, preferences for location such as vertical and horizontal distribution, envelope criteria (second skin), and atria were taken into account. The architect had to dialog with the model in real-time by interacting with the room elements to generate massing scenarios, creating time and cost efficiencies while also revealing space-left-over-after-planning (SLOAP) opportunities (Figures 4.2 and 4.3).

Models of Occupant as User

It would be a fallacy to believe that we can simulate the occupant, but there are properties within geometric configurations that allow an approximation of the potential affordances for actions that an occupant perceives. It can be assumed, however, that generic behaviors can be inferred from spatial configurations and vice versa that spatial conditions can be inferred from occupant actions. These generic actions from cognitive interpretations of spatial conditions can be regarded as the occupant's heuristic, such as way-finding principles [8]. As Lynch demonstrated through mental maps [4], standard architectural representation, like metric plans, is not necessarily how an occupant perceives his or her field. For user-centric design based on cognitive properties, metaspatial maps must be compiled that decode the bodily experience of the occupant. This correlation between the computational model and the experience of the user in the field is what Papert called body-syntonicity.

Relational Representation

Syntactic representations for relations between occupation and geometry, theoretically generalized for all spatial plans, can be found in Michael Benedikt's isovist [9] or Hillier's j-graphs [10]. Benedikt's isovist is a geometric representation of a location's viewshed in plan (assuming a 360-degree vision). Extracting the geometric boundaries of the viewshed, spatial aspects can be analyzed, which inform movement behaviors such as long vistas (desire lines). Hillier's justified graphs (j-graphs) use a dual-graph representation to map the topological connectivity of permeable space partitions. The graph can be adjusted for each partition within a building plan and therefore shows how buildings contain many permeability configurations. These graphs can be evaluated topologically, indicating likelihood of allocations of social and operational functions or movement affordances. Benedikt's isovists can be automated across all visible edges to produce visibility graphs for further processing of movement networks and visual connectivity.

Just from this limited set of configurational representations, a whole host of metaspatial maps can be produced that relate to cognitive properties of places. CDR extended Benedikt's and Hillier's planar representations into volumetric models for the spatial and perceptive analysis of the National September 11 Memorial Museum in New York [11]. The viewshed was projected into space and onto all surfaces, which allowed for a visual integration measure of the volumetric building interior. Visual integration shows which parts of space are most visible within a building. The museum provides a descending route between former north and south towers, creating a narrative rhythm of movement by allowing glimpses of

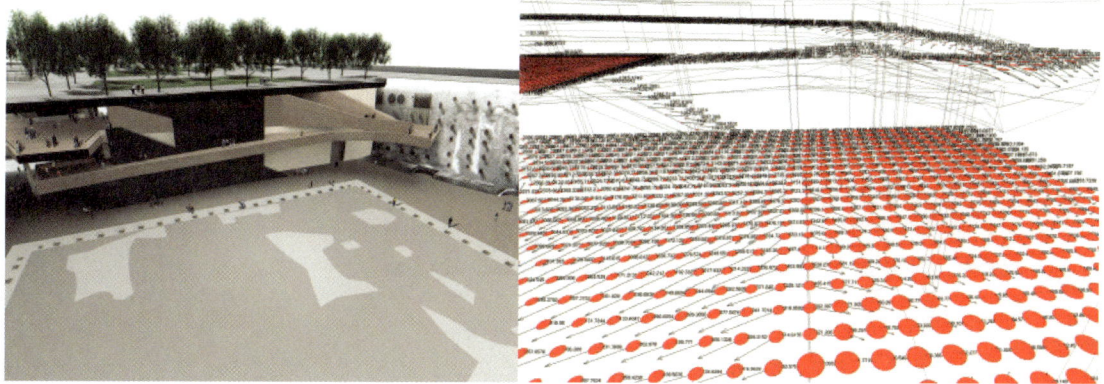

Figure 4.4   National September 11 Memorial Museum, New York, 2007. Left: Section through the underground museum showing the descending ramp through the former North Tower. (Copyright Davis Brody Bond architects and planners.) Right: Visual analysis of qualities of views and their attributes across walkable surfaces of a ramp and the North Tower. (Copyright Aedas architects.)

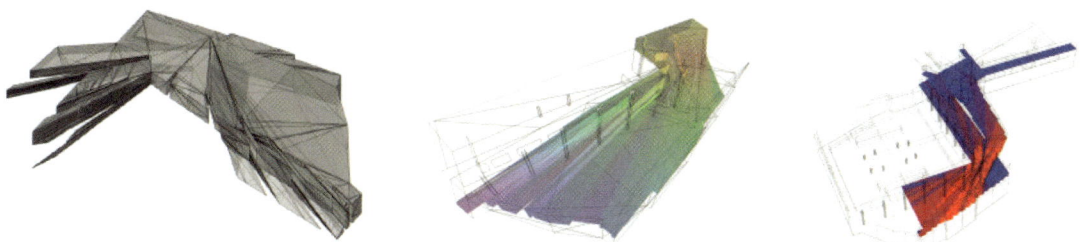

Figure 4.5   National September 11 Memorial Museum, New York, 2007. Three visibility analysis types. From left to right: Single 3-D viewshed geometry, instance of view along the ramp, and horizontal section through movement along the ramp through museum. (Copyright Aedas architects.)

strategically placed artifacts. The volumetric isovist analysis we developed evaluated whether the glimpses occurred at the desired strategic locations and where to place artifacts in space so that they connect a series of locations simultaneously. Subsequently, a 4-D visual analysis was developed where viewshed geometries can be analyzed along routes over time (Figures 4.4 and 4.5).

Enactive Architectures

The intrinsic correlation between space and use is akin to enaction theory, where knowledge requires action. In architecture and cognitive theory, interaction with the environment is considered key to learn intuitive spatial thinking. Modernists optimized configurations by designing out interaction between people and with places. In reaction to this social engineering of space, organic architects in Europe created theories of bodily architectures (Figure 4.6).

Haering's concept of wesenhafte Gestalt provided the theoretical basis for the German organicists [12]. A form generation process is intended where spatial configurations emerge from the interaction between human behaviors and surrounding space. Wesenhaft refers to the essence of a person or structure and thus indicates that spatial structures (Gestalt) emerge from a kind of mapping of user behaviors, implying that form reflects more purpose than a simple resolution of functional efficiency.

Similarly, psychologist James Gibson elaborated an ecological theory

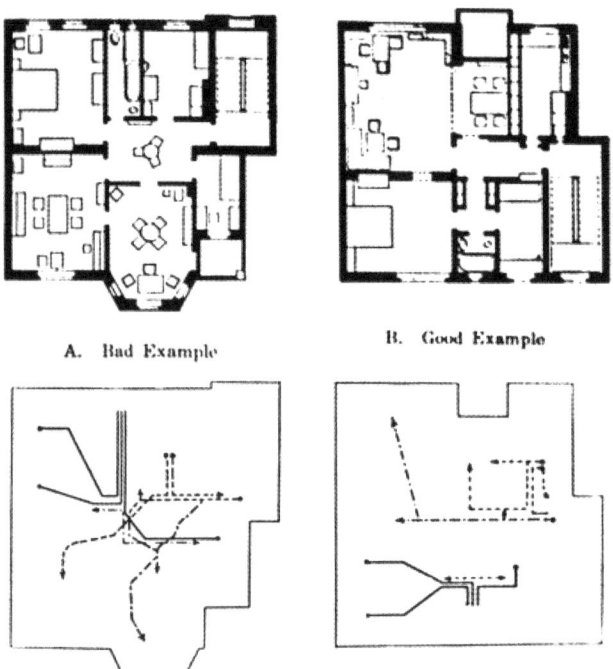

Figure 4.6 Alexander Klein, 1928: The Functional House for Frictionless Movement. (Copyright: found in Robin Evans: *Translations from Drawings to Buildings and Other Essays*, AA Documents 2, AA Publishing, London, 1997.)

of perception, which required the user to move and occupy in order to perceive [8]. Gibson believed that cognitive affordances are encoded in the spatial environment and are decoded by the user through movement and vision.

We tested this type of correlation for a feasibility study for the remodeling of the public realm in front of Euston station in London, 2013. A design simulation was created that synthesizes route and visibility analysis with morphology drivers for building scales and area development quantum. Entrances to a new proposed massing on Euston Square were mapped into the contextual street network calculating which routes would be used to access the new station massing. Street sections along those routes were analyzed to compile a value for experience of scale for each entrance according to the routes that those entrances might activate. Additionally, a visual choice map provided information for all entrance locations in terms of orientation when exiting. Visual choice relates to the number of urban spaces at a location that a user could take.

In the interactive model, the designer can place, move, and delete entrance/exit locations along a prespecified massing envelope. Depending on number and configuration of points, the massing is adjusted automatically to provide a coherent experience for the user when either accessing the station along one of the calculated routes or exiting the station into one of the urban places. The context is mapped into the building massing, resulting in gross-floor area recommendations and morphological variation correlating to the contextual experience (Figures 4.7 and 4.8).

Figure 4.7 Euston Station, forecourt feasibility study, London, 2013. Left: Visual impact of a massing option from site routes. Right: View from inside to forecourt. (Copyright Aedas architects.)

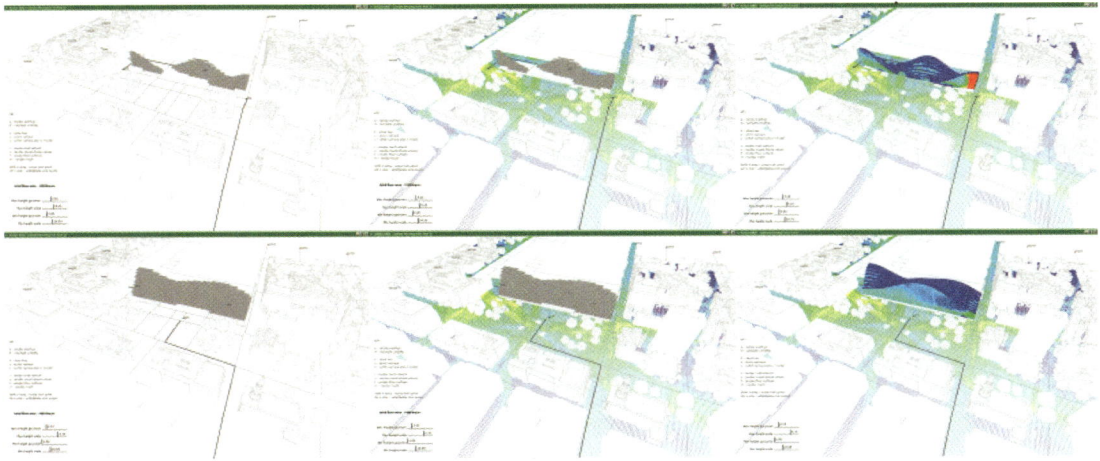

Figure 4.8 Euston Station, forecourt feasibility study, London, 2013: Integration of generative and analytical massing showing two sequences of floorplate design from access routes, entrance definition from visual qualities, and visual impact of massing on site. (Copyright Aedas architects.)

Generic Functions of Buildings

The German organicists called this correlation concordance (Konkordanz), where occupation and spatial configuration are mutually supportive. It hints at movement and general occupation of the building as the essential determinant that lends a building or place its function. Hillier called this the generic function of buildings that can be approximated via spatial analysis of the permeability structure. Hillier in analysis and German organicists like Scharoun via their design process assume that the operational function is of a specific building typology is so strongly programmed and allocated in bespoke areas (rooms) that it does not affect the overall functioning (use) of the building. Inversely, circulation and semipublic areas connecting functional areas represent the actual performance of a building/place, which are ironically unspecified in most design guidance other than the above-mentioned discursive implicit aspects.

In 2011 CDR did a behavioral mapping project for the Polish Embassy in London to visualize the inverse relation between the consulate's functional operation, generic occupation, and its spatial layout. A simulation was produced to help configure the interface between the consulate's public operations (passport, payments, visa applications, etc.) with the customers' use of space. The use of the existing consulate was mapped and the spatial

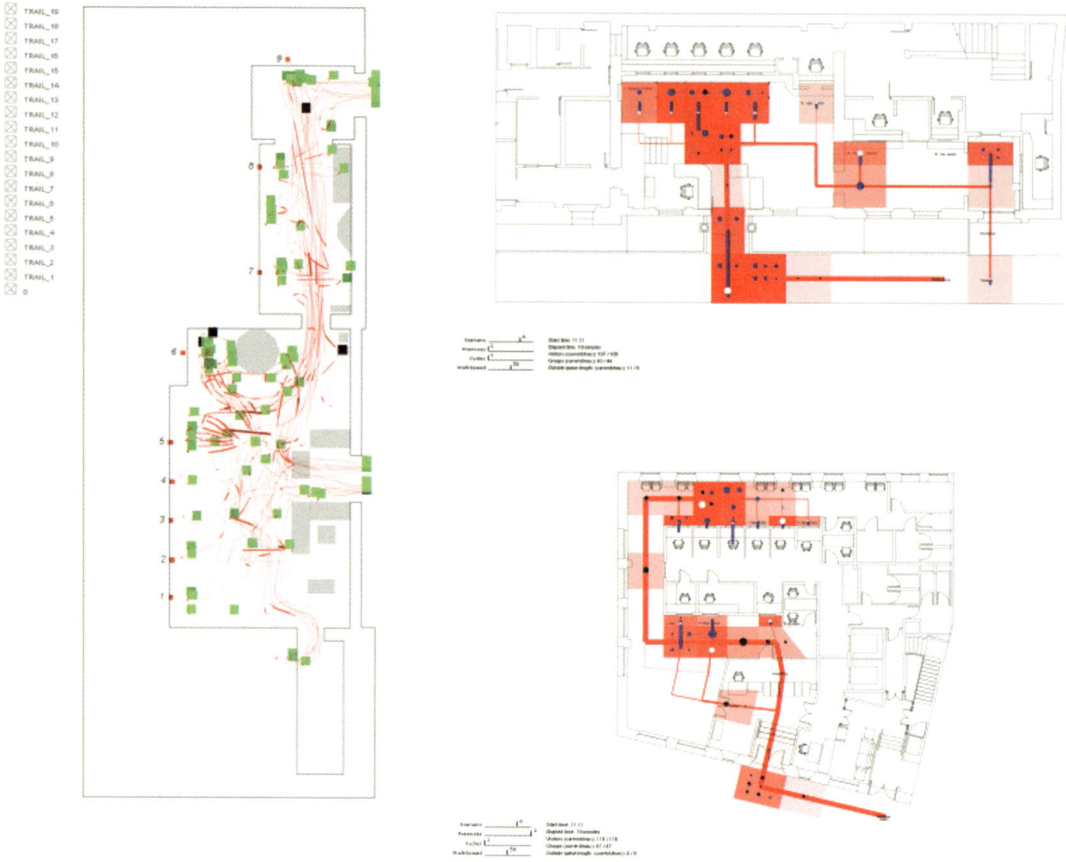

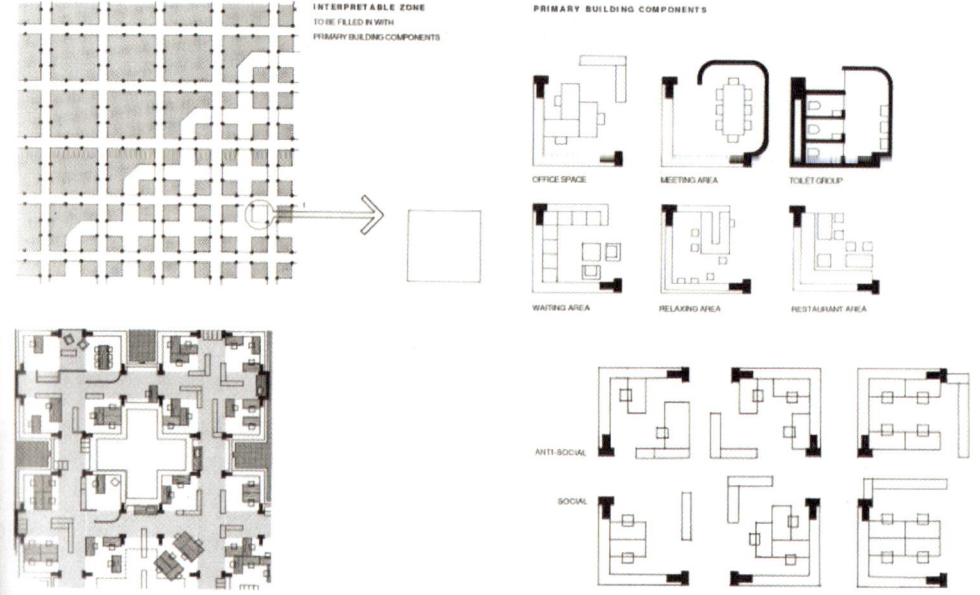

Figure 4.9 Left: Polish Embassy, operational mapping, London, 2011: Movement patterns from spatial arrangements of operations. (Copyright Aedas architects.) Right: Polish Embassy, operational mapping, London, 2011: Simulation of occupation patterns and operations schedules on existing premises and projected into planned premises. (Copyright: Aedas architects.)

Figure 4.10 Herman Hertzberger, Centraal Beheer Building, Apeldoorn, 1972: Plan modules and their affordances for occupation are aggregated into global patterns for polyvalent use. (Copyright Architektuurstudio Herman Hertzberger.)

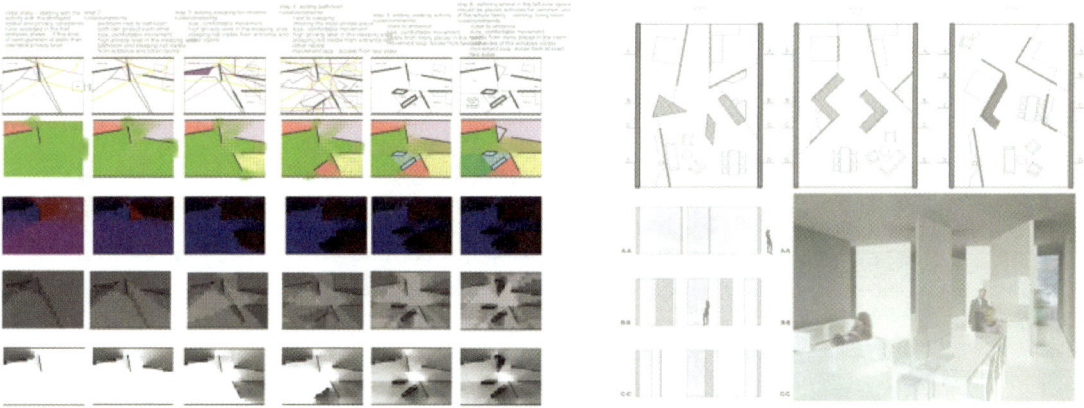

Figure 4.11 Floating Room project, Technical University Munich, 2012. (Copyright: Christian Derix.)

statistics of operation and occupation visualized in scenario simulations. The new consulate layout is now simulated with various mapped patterns of occupation and schedules of operations projecting forward possible generic functions (Figure 4.9).

Designing Organically

Designing organically doesn't mean to create shapes that look like complex biological structures; it means to provide affordances for organic use. Ego- and body-syntonicity need to be synthesized into a design process, overcoming the isolation between configurational analysis and heuristic generation, so that concordances in the design process can be approximated. Hertzberger has provided successful examples of such organic design systems even without computation, such as the Centraal Beheer offices. A grammar of potential social and operational occurrences is developed through an alphabet of local structures that are progressively aggregated into global spatial configurations within a framework grid. The resulting generic function of the building emerges from the aggregation and is not preplanned other than main flows for building code constraints. This approximation of global configuration by assembling local affordances is what Scharoun called "improvisations." It purposefully neglects a global formal appearance of a building but concentrates on the correlation between cognitive affordances and generic performances. Improvisations leave space for emergent occupation [14], created from intersections of polyvalent local structures: where the designed ego-syntonic components interface, new body-syntonic actions can occur (Figure 4.10).

During a visiting professorship at the Technical University Munich in 2011/12, we conducted an ambitious design studio project where students had to learn how to observe and map spatial phenomena in order to extract rules of spatial dynamics. Measures from subjective perceptions had to be generalized to establish computable rules representing the correlation of use and space. Those led to a hybrid design system where analog physical models and computational heuristics were mixed to generate spatial configurations. The hybrid system had to be organized as an iterative workflow where computation and experience weigh each other rather than separating analysis and generation. Six such systems dealing with different typologies were developed, such as the Floating Room project developing residential apartment layouts [15]. The premise of Floating Room stated

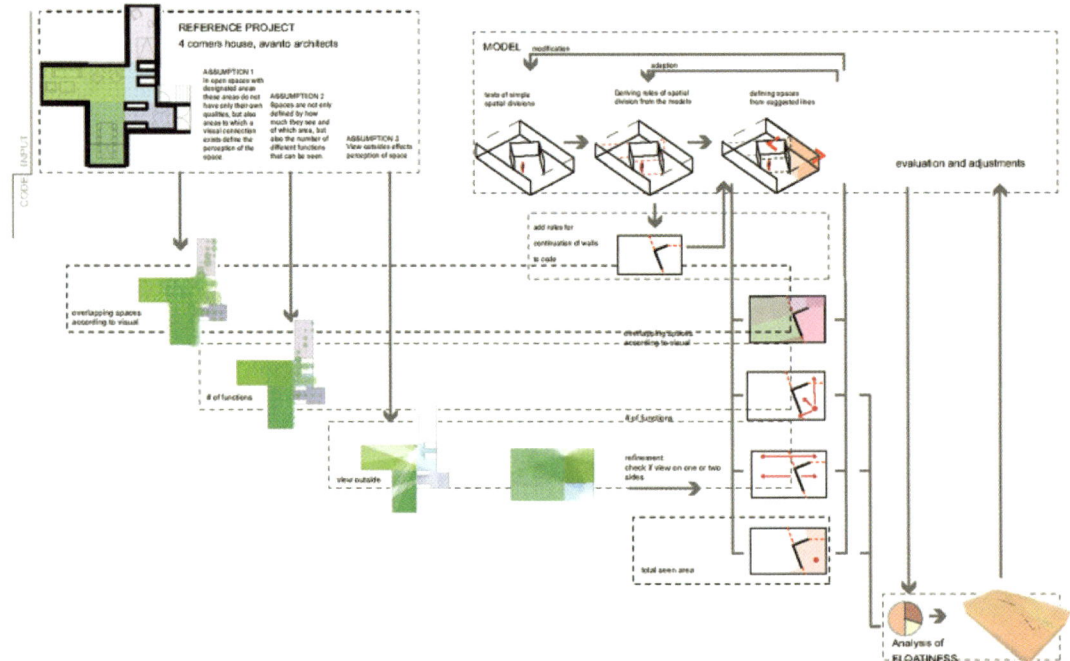

Figure 4.12 Floating Room project, TUM, Munich, 2012. (Copyright Christian Derix.)

that physical spatial delineations are vague even on small scales and building plans should be assembled by simulating local experiences. Physical models were used to validate spatial rules, and computation was used to evaluate implicit qualities like privacy criteria, room usages, and external connections. Multiple grid-based graph values for visual integration and topological connectivity were developed to evaluate local experience. The geometry of the layout was aggregated incrementally employing hybrid rules. Context and order of aggregation led toward final apartment layouts within a residential block in Helsinki.

Computational Archetypes of Space

The computational design community generally mistakes architecture with objects. Architecture is meant to provide experiences by using spaces. We differentiate buildings by experience, not by construction or shape. We believe computation is a notational system that allows us to approximate experiential archetypes. Buildings and urban spaces are defined by their ego- and body-syntonicity, not classified simply by functional areas and specifications as approached in the traditional sector-based understanding. Archetypes are defined through affordances of cognition and action that we can approximate computationally. Hence spaces are not classifiable as prototypes as in industrial design but as use and configuration archetypes. In order to design with the correlation between space and use algorithmically, computational designers need to become more architectural and observe the heuristics of places, spaces, and users.

References

[1] Rittel, H., and Webber, M. "Dilemmas in general theory of planning." *Policy Sciences*, Vol. 4, pp. 155–169, 1973.

[2] Allen, S. *Practice: Architecture Technique + Representation*, 2nd Edition. Routledge, New York, 2009.

[3] Commission for Architecture and the Built Environment (CABE): ByDesign. DETR, London, 2000.

[4] Lynch, K. *The Image of the City*. MIT Press, Cambridge MA, 1969.

[5] Otto, F. and Rasch, B. *Finding Form: Towards an Architecture of the Minimal*. Edition Axel Menges, Stuttgart, 1996.

[6] Papert, S. "The Mathematical Unconscious." In Wechsler, J, *On Aesthetics in Science*, MIT Press, Cambridge MA, 1981.

[7] Derix, C., and Izaki, A. "Spatial computing for the new organic." In *Computation Works:The Building of Algorithmic Thought*, DeKestelier, X. and Peters, B. (eds.), Architectural Design (AD), Boston–London, 2013.

[8] Passini, R. "Wayfinding design: Logic, application and some thoughts on universality." *Design Studies*, Vol. 17, No. 3, July 1996.

[9] Benedikt, M. L. "To take hold of space: Isovists and isovist fields." *Environment and Planning B*, Vol. 6, pp. 47–65, 1979.

[10] Hillier, B. *Space is the Machine*. Cambridge University Press, Cambridge, 1996.

[11] Derix, C., Gamlesaeter, A., and Miranda, P. "3D Isovists and spatial sensation: Two methods and a case study." In *Movement and Orientation in Built Environments: Evaluating Design Rationale and User Cognition*, Haq S., Hölscher, C. and Torgrude, S. (eds.), EDRAMove conference, Veracruz, 2008.

[12] Janofske, E. *Architektur-Raeume—Idee und Gestalt bei Hans Scharoun*. Vieweg & Sohn Verlagsgesellschaft, Wiesbaden, 1984.

[13] Gibson, J. J. *The Ecological Approach to Visual Perception*. Houghton Mifflin, Boston, 1986.

[14] Hertzberger, H. *Lessons for Students in Architecture*. 010 Publisher, Rotterdam, 2005.

[15] Derix, C. "Implicit Space." *Proceedings of the 30th European Computer Aided Architectural Design in Education (eCAADe)*, Prague, 2012.

Chapter 5

Algorithmic Principles for Façade and Building Automation Systems: Al-Bahar Towers, Abu Dhabi

Abdulmajid Karanouh

Introduction

The Abu Dhabi Investment Council New Headquarters (Al-Bahar ICHQ Towers, Figure 5.1) is an international competition won by Aedas Architects in Collaboration with Arup in 2007. The office twin towers stand at 150 m high in Abu Dhabi in the United Arab Emirates in the Gulf Region.

Islamic and regional architecture, sustainable technology, and inspiration from nature form the foundation of the design concept of Al-Bahar Towers (Figure 5.2). Among its many performance-oriented design features, the building mainly stands out with its fluid form, honeycomb structure, and dynamic shading screen that adapts to the changing environment, offering the building a unique identity.

The building won Best Innovation Award for the year 2012, and was offered by the Council for Tall Buildings and Urban Habitat (CTBUH) among many others. The concept and design development of the building offer a model of how design teams may respond to recent calls for more sustainable building design and to local initiatives like Estidama 2030. It also introduces hybrid systems and technologies inspired from nature

Figure 5.1   Al Bahar Towers: a natural outcome of contextually inspired and performance-driven design.

and the past and reintroduced in a contemporary manner to relate to the context and to address the client's aspirations.

In order to realize such an ambitious and highly complex design within a relatively constrained budget and program, a special design-to-delivery strategy had to be developed for this purpose. A series of bespoke algorithms were developed following underlying mathematical principles inspired from the universal order of orbital motion. A special data communication mechanism was developed in order to convey these algorithms into a construction-industry-friendly format that the supply chain could understand and deliver.

Key Design Elements

Geometric composition: The intersection of the infinite arrangements and populations of circles (2-D) and spheres (3-D) generate infinite arrangements of nodes that—when connected following a mathematical logic—generate the famous Islamic patterns and forms. However, the geometric composition principles chosen for the Al-Bahar Towers are derived from hexagonal triangulation (Figure 5.3).

Main structure: An intelligent honeycomb formation inspired from beehives and uniquely applied to towers provides a highly efficient and robust system.

Main form: All floor-plates and vertical profiles of each tower are made of tangential arcs following specially devised mathematical rules. They generate an intelligent fluid form that offers orientation while maximizing volume to envelope ratio, admission of natural diffused light and distribution, and minimizing wind-load impact on the building skin and supporting structure.

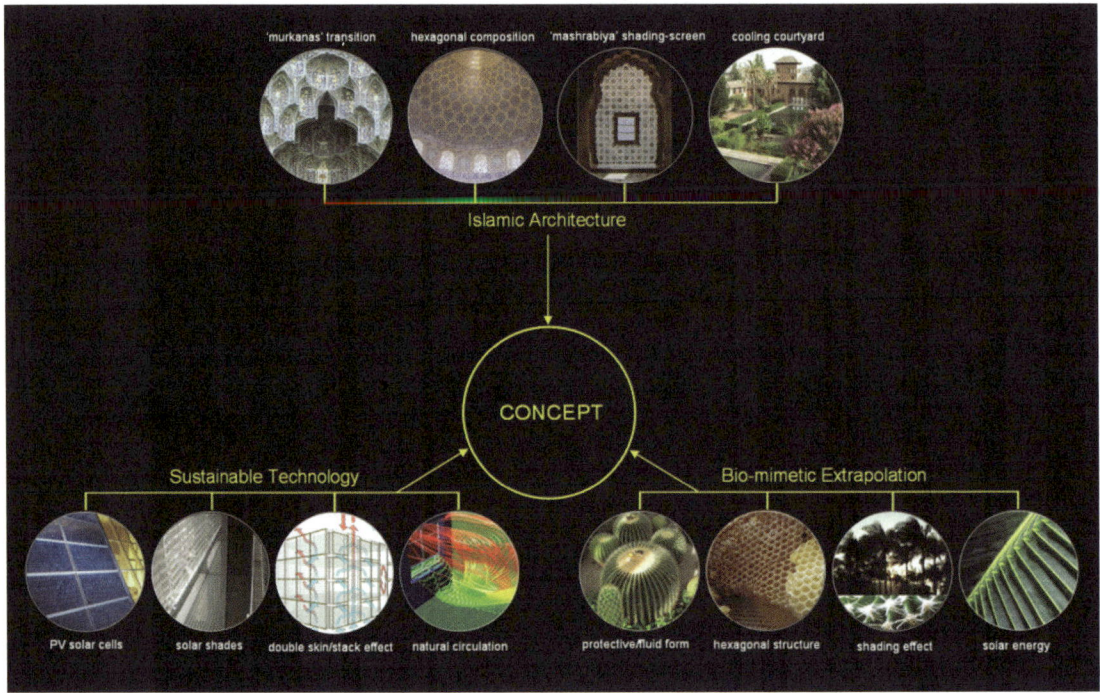

Figure 5.2 ADIC concept and philosophy displaying the key design elements of each of the main concepts.

Figure 5.3   Circles and spheres form the base of Islamic geometric composition.

Façade: Over 1,049 mechanized folding components per tower form the external adaptive shading screening to the weather-tight glass skin to protect it from excessive solar gain (Figure 5.5).

The design brief was based on the aspiration of having a building that represents the ethos of the Investment Council while relating to the context and the underlying cultural tradition of Abu Dhabi.

Adaptive Mashrabiya Solar Screen

The main standout feature of the Al-Bahar Towers is the adaptive Mashrabiya; a unique kinetic shading screen that comprises triangular units that fold/unfold like umbrellas at various angles offering louver-like geometries in various positions.

Concept: The design concept was originally inspired from the Mashrabiya —a traditional lattice shading screen particular to the Middle East—that reduces direct solar rays reaching the main skin of the building. Additional inspiration was drawn from natural systems like leaves and flowers that adapt to the moving sun (Figure 5.4) to improve efficiency, and thus the dynamic Mashrabiya was born. The screen will reduce solar glare while providing better visibility by avoiding dark tinted glass and internal blinds that distort the appearance of the surrounding view. This sophisticated system offers better distribution of natural diffused light and optimizes the use of artificial lighting through dimmers linked to light perimeter sensors. The system also reduces requirements for air-cooling loads due to reduction of solar gain of the main skin. The system will help in reducing the overall energy consumption, carbon emission, and plant room size. The flexible smart folding geometry was carefully worked out to overcome the limitations of traditional vertical and horizontal louvers especially when applied to geometrically complex buildings.

Distribution: There are 1,049 units fitted to each of the towers covering the east, south, and west zones, leaving the north face exposed (Figure 5.5) where there is no exposure to direct sunlight. When a façade zone is

Figure 5.4 Concept of the shading screen inspired from the Mashrabiya and natural adaptive systems.

subjected to direct sunlight, the Mashrabiya units in that zone will deploy into their unfolded closed state providing shading to the inner glazing skin. As the sun moves around the building each Mashrabiya unit will progressively open (Figure 5.5).

Control software and building management system: During the competition stage, the Aedas R&D team collaborated to produce a bespoke program using a Java stand-alone applet that simulated the path of the sun and the kinematics reaction of the shading units. Human-machine interface software was later developed using Siemens and SCADA control systems (Figure 5.19).

Managing Complexity

Each tower comprises over 3,000 unitized curtain-wall panels and 1,049 Mashrabiya units. Similarly, the main structure comprises thousands of components of unique sizes. This poses a great challenge in terms of managing complexity during the design development stage all the way to fabrication/manufacturing and construction. This challenge was tackled as follows:

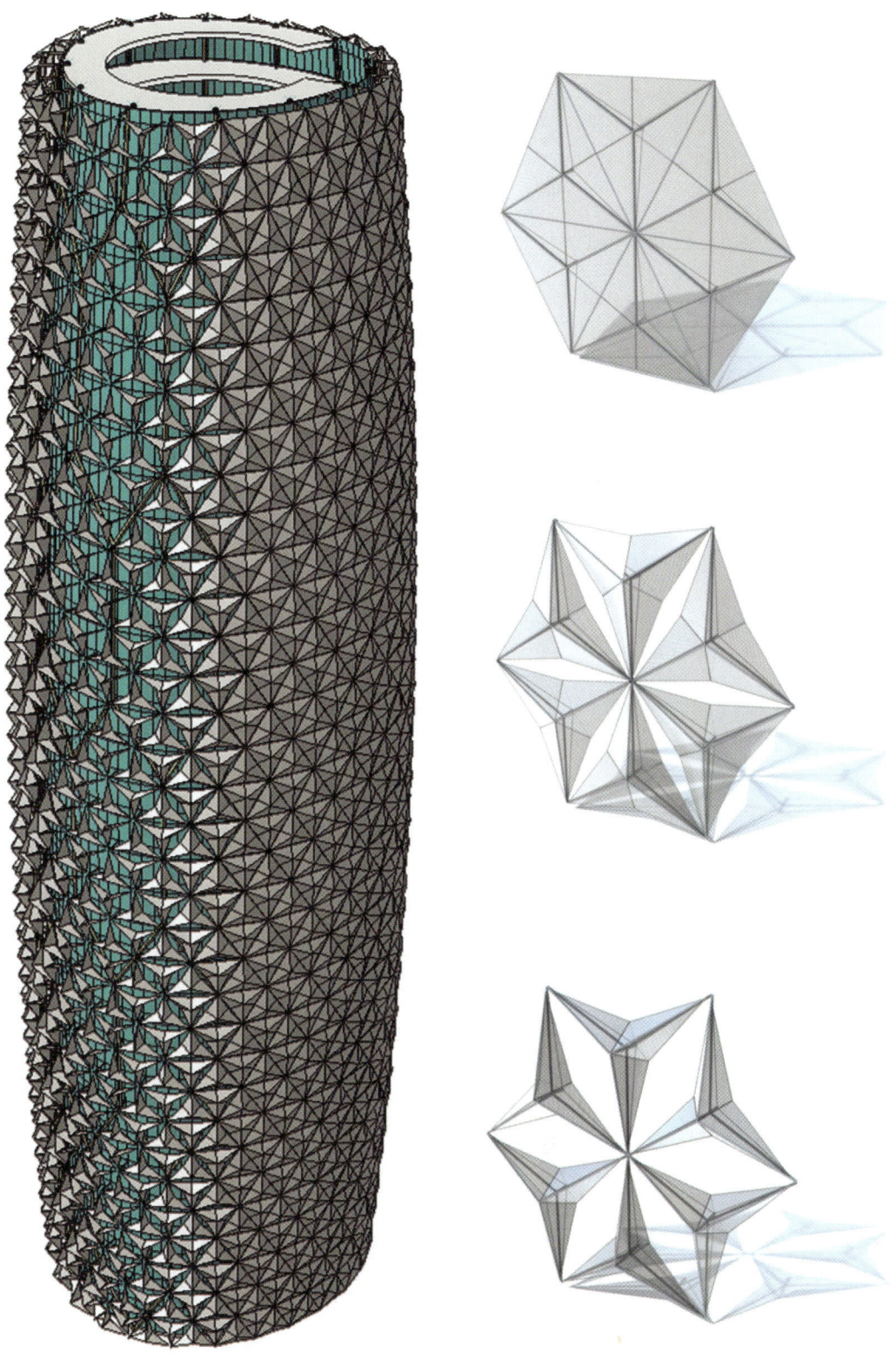

Figure 5.5 Top to bottom (right): Six Mashrabiya units at unfolded, mid-folded, and max-folded positions.

Algorithmic thinking: Designing by explicit rules where the geometry and performance criteria are both mathematically and technically pre-rationalized meant that at no time the process would/should ever be locked down to a single CAD/BIM package or 3-D model. This offered every related party (architect, engineer, contractor, subcontractor, etc.) the freedom to use whatever tools they were comfortable with as long as they followed the preset rules and produced their scope within the specified tolerances.

Universal solution: The complex nature of the geometry of the building and behavior of the mechanized units as described earlier meant that components would connect and take positions in numerous different angles and coordinates. It was therefore very important to adopt a universal design approach where connections and interfaces would be designed to accommodate and adjust to most different cases and scenarios. This would therefore limit the complexity associated with unique sizes while minimizing the number of unique engineered solutions.

Computation: The above lends itself quite well to parametric design enabling the use of different types of tools efficiently. This included using various packages like Grasshopper, Digital Project (CATIA), Tekla, Inventor, and the use of SolidWorks, among many others. This allowed direct data extraction from the digital models and feed into CNC machines for fabrication and then into topographic survey machines on site for installation. Over 15 different software and packages were used by various parties to develop and deliver their scopes.

Communication: The major challenge, however, on a project of this scale and complexity is how to communicate an algorithm to a project team and supply chain that do not necessarily have the required background and level of understanding in terms of mathematics and programming. Simply handing over a computer code won't be enough. A special and rather unique geometry construction and performance manual was developed for this project, which drew inspiration from nature's DNA (set of instructions of what and how to build and function a human body, for example) and from LEGO toy manuals, which will be elaborated on in later chapters.

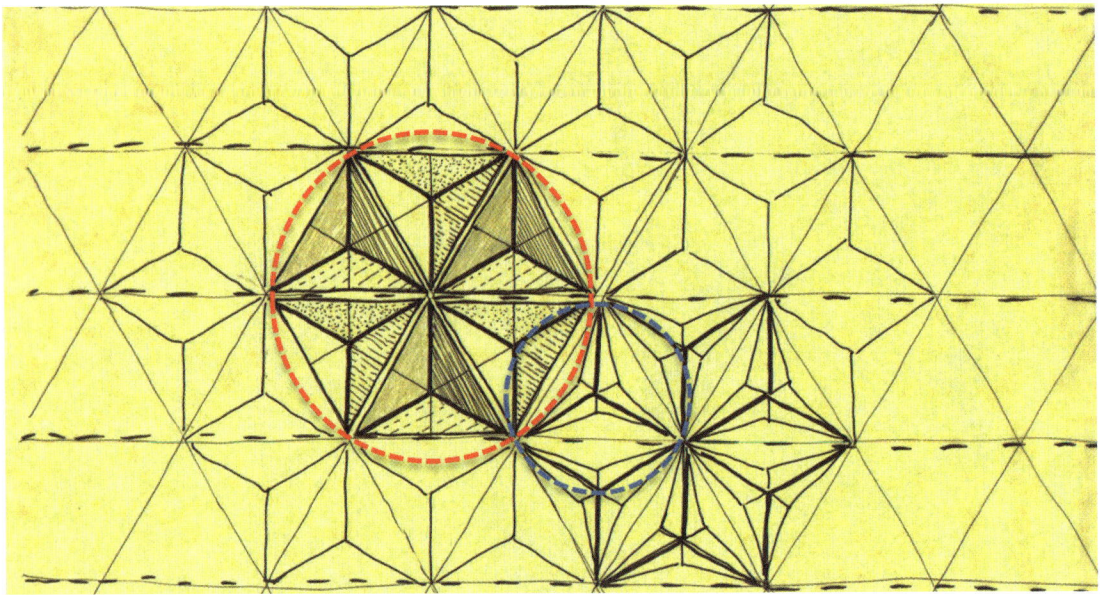

Figure 5.6   First freehand sketch of the first Mashrabiya adaptive solar screen intent.

The Algorithm

The mathematical principles behind the algorithm that generates the geometry of the buildings' components are relatively straightforward and consistent across the various items of the towers. All main components are constructed from a series of circles and spheres as described earlier. All mechanical components' motions are traced by a combination of circular and spherical motions.

The focus of this chapter however will revolve around the adaptive Mashrabiya solar screen. A series of exercises had to be performed prior and after writing the first algorithm for generating the geometry and operation of the Mashrabiya adaptive solar screen in order to validate it and later communicate it to all other related parties. The development process of the algorithm underwent the following stages:

- Freehand sketches (Figure 5.6)
- Physical paper models (origami, Figure 5.7)
- Performance criteria (Figure 5.8)
- 2-D CAD geometry construction
- 3-D CAD geometry construction
- Setting up first set of algorithmic principles
- 4-D Java scripting
- 4-D parametric/BIM model
- Geometry optimization

Figure 5.7 First origami physical model and 4-D simulation model of the folding/unfolding mechanism.

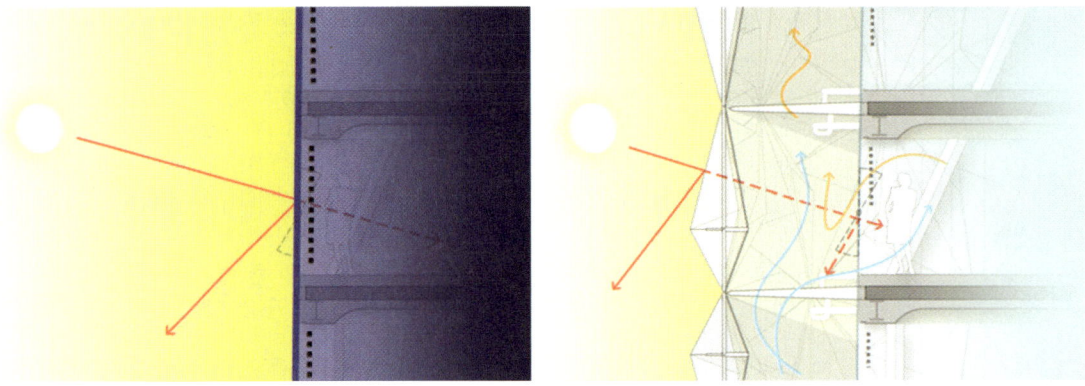

Figure 5.8 Left: Without Mashrabiya (static solution). Right: With Mashrabiya (adaptive solution).

- Solar study
- System optimization
- Geometry construction and performance manual
- Updating the adaptive algorithmic principles
- Writing the code for the final HMI control software

Setting Up First Set of Algorithmic Principles and 4-D Java Scripting

Right after setting up the first 3-D-model of the Mashrabiya origami paper in Rhino, a few principal experiments and assumptions were made as to how the Mashrabiya would move and arrange itself around the building. There was a need, however, to populate the Mashrabiya units en mass around the towers and get them running in real-time. This task was given to Pablo Miranda—architect and computational designer and former member of the R&D Computational Design Group at Aedas. The principles of the algorithm were agreed on with Pablo, who went on to develop a program using Java Script (Figures 5.9–5.16).

Below is a description of the process in Pablo's own words:

"The intention of the program for the visualization of the Mashrabiya actuated façade was to provide a fast way of generating an accurate geometric representation of the elements that constitute it (at a schematic level), as well as evaluating its possible responses to different sun angles during the different days of the year. The sequence of the program is as follows: first, a triangular mesh is read by the program; this mesh represents the position and shape of all the triangular façade components. The program then associates a Mashrabiya element to each of these triangles. Each of these elements has a description of all the geometrical relations of the Mashrabiya and is capable of calculating through a simple formula (which is the result of the development of the equations representing these relations and constraints, code snippet 1) the shape of the Mashrabiya for a specific position of the linear actuator placed in the center of the component, or in other words, it is capable of calculating the shape of the Mashrabiya for any given aperture angle. This is done in a sort of generic coordinates (identical to all components), and later transformed into the specific triangle of the façade using barycentric coordinates (code snippet 2). The process assumes that the initial mesh is correct (all triangles are equilateral). In the case this is not true, the

Figure 5.9　Java scripting.

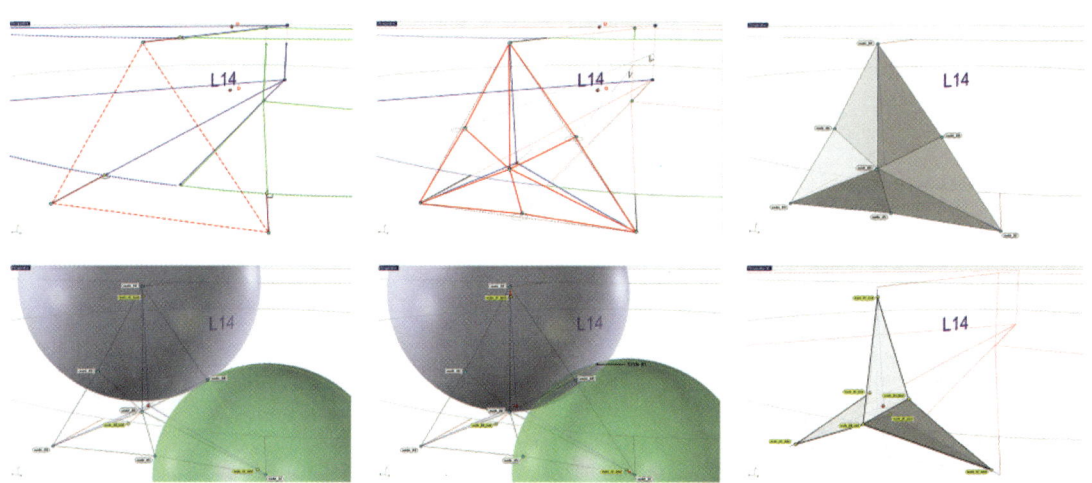

Figure 5.10　Java scripting.

program will work, but the produced geometry will be distorted according to the triangles. The application provides a simple user interface (through a slider) for modifying the position of the linear actuators and generating and evaluating the corresponding resulting geometries as well as saving these geometries in DXF format.

In a further variation of the program, the aperture of each of the elements of the Mashrabiya is made to correspond to the incidence angle of the sun. Depending on the angle of the normal triangular face defining on each element with the sun angle (calculated through a simple formula for different year, month, day, and time), each element will more or less unfold. When the angle is small, that is, when the sun rays incident is straight and orthogonal to the element, it unfolds to prevent direct sunlight and excessive glare into the building. On the other hand when the angle is high (at noon for example) the element folds, providing shade and offering better views. The program provides an interface to set the month of the year calculating and visualizing in real time how the Mashrabiya performs throughout the day. The interface offers the possibility to tweak a number of parameters (such as setting the minimum and maximum aperture angles for all elements) and to freeze and capture the geometry at any time."

```
private void calcGeometry()
    {

        height=pos+centreZOffset;

        length=(float)Math.sqrt(longSide*longSide-
height*height);

        float vx= length/longSide;
        float vz=-height/longSide;

        /**
         * we find the line that passes through h, is in the
folding plane (on s1)
         * and has an angle with the h-1 line that is 60. This
results in a quadric
         * equation.
         */

        float a= vz*vz*(1+tan60*tan60)+vx*vx;
        float b = -2*cosAlfa*vx;
        float c = cosAlfa*cosAlfa-vz*vz;

        float sqint=b*b-4*a*c;
        sqint=sqint<0? 0:sqint;

        float ux=(float)((-b-Math.sqrt(sqint))/(2*a));
        float uy= ux*tan60;
        float uz= (cosAlfa-vx*ux)/vz;

        pts[4][0]=ux*shortSide;
        pts[4][1]=uy*shortSide;
        pts[4][2]=vz>=-0.005 ? -pfZOffset :
uz*shortSide+height;
    }
```

Figure 5.11 Code snippet 1: Geometric formula for the Mashrabiya.

The formula solves only one point (the position of one of the straight angles of the triangular pieces of the element). All other points can be deduced from this or through simple trigonometric relations.

```java
private void calcBaryCoor(float[] p, float[] bar)
{
    /**
     * BaryCentric coordinates formula for a triangle
     * from:
     *http://en.wikipedia.org/wiki/Barycentric_coordina
     tes_(mathematics)
     */

    //the triangle is defined by pts[1], pts[4] and pts[6],
    //the vertices of the triangle.

    float a=pt1[0]-pt3[0];
    float b=pt2[0]-pt3[0];
    float c=pt3[0]-p[0];

    float d=pt1[1]-pt3[1];
    float e=pt2[1]-pt3[1];
    float f=pt3[1]-p[1];

    //this is according to formula...
    //since we are using a flat triangle we should not do this, and
    //make them all 0 instead

    bar[0]=(b*f-c*e)/(a*e-b*d);
    bar[1]=(a*f-c*d)/(b*d-a*e);
    bar[2]=1-bar[0]-bar[1];
}
```

Figure 5.12 Code snippet 2: Baricentric coordinates. (Image courtesy of Pablo Miranda, 2009.)

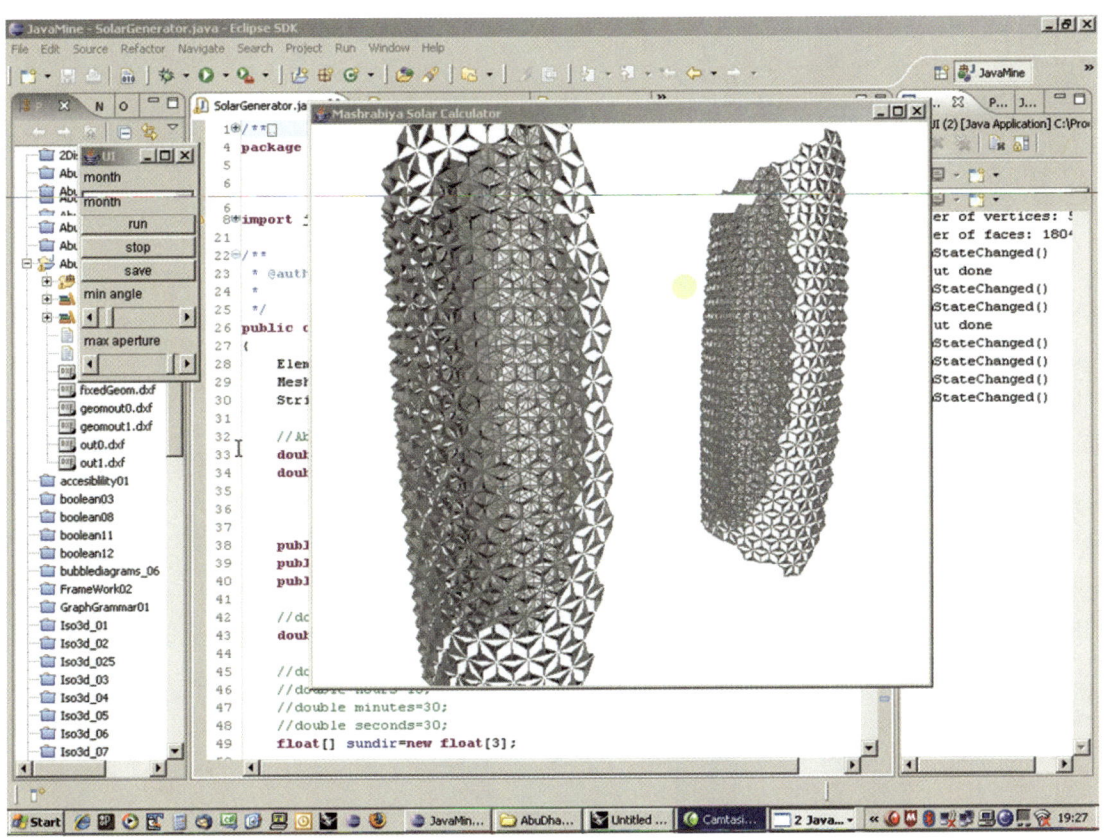

Figure 5.13 A Java bespoke control program was developed during the competition stage by architect and computational designer Pablo Miranda simulating the motion of the sun and the reaction of the shading screen components to it.

4-D Parametric/BIM Model and Geometry Optimization

After the preliminary geometric construction, kinetic motion, and adaptive algorithmic principles were set for the Mashrabiya shading screen, it was time to analyze the preliminary assumptions made earlier in the competition and produce a proof of concept. This task required a sophisticated mathematical solver. The project team hired Gehry Technologies [GT] for this task as they can operate and customize advanced platforms like Digital Project—a CATIA based tool common in the automotive and aerospace industries.

The budget for hiring specialists was limited, therefore GT's scope had to be carefully formulated and very specific. The author teamed up with Andrew Witt—former director of design innovation at Gehry Technologies—to develop the system. Andrew is trained as a mathematician, designer, and programmer, and is interested in the application of rigorous mathematical approaches to solve complex problems—a perfect partner for this task. Below is the description of the process of implementing BIM in Andrew's own words:

"The Al-Bahar Tower's project presented several remarkable opportunities for parametric design, nonlinear optimization, and kinematic simulation, particularly in the mechanical design of the sophisticated Mashrabiya system. Over a period of several months I participated as a Gehry Technologies consultant to the leading designer, Abdulmajid Karanouh.

"We used several advanced methods including functional optimization, constraint-based kinematics systems, and stochastic simulation to understand a series of fascinating implicitly defined puzzles surrounding the dynamic behavior of the building. GT's engagement on the project focused on building a BIM reusable for construction and a capacity to model and optimize implicitly defined nonlinear problems which may influence the environmental performance of the building and the mass customization of its components.

"We began with a full parameterization of the model, which allowed us to compare various stochastically optimized global forms for their relative envelope area to volume ratio, and thus their environmental performance. In order to truly model the daily solar gain of the building, it became necessary to understand and model the dynamic behavior of the hundreds of Mashrabiya covering the façade. In particular, we wanted to model an optimal folding function, so that we could shade as much of the building as possible while obstructing views as little as necessary. It quickly became clear that an idealized, origami concept of the Mashrabiya could not adequately represent the true dynamic behavior of these elements. The thicknesses of the joints, hinges, and materials created small but significant divergences from the idealized motion model. The compound motions were highly nonlinear, and thus we modeled them with an implicit kinematics solver. But the kinematics themselves were also constrained by extrakinematic factors, in particular solar motion. By integrating all of this into one model, we were able to fully control geometry, mechanics, and functional outputs in a dynamic way from a single performance constraint: solar gain (Figures 5.13—5.21).

"What became interesting about the process was the very mechanical behavior of the joints and actuators had an emergent impact on the solar performance of the building. Thus the optimization of the nonlinear function of solar shading created implicit constraints on the mechanics of the Mashrabiya, itself a highly nonlinear problem. It was a classic inverse function boundary value problem, but one which was conceptualized as a combination of geometric and analytic terms, not simply functional ones.

"It was an extremely interesting process because Abdulmajid had defined the geometric conception of the project so clearly from the beginning; it lent itself perfectly to a parametric approach. It also presented an unusual opportunity to think about performance-based kinematics. Probably the

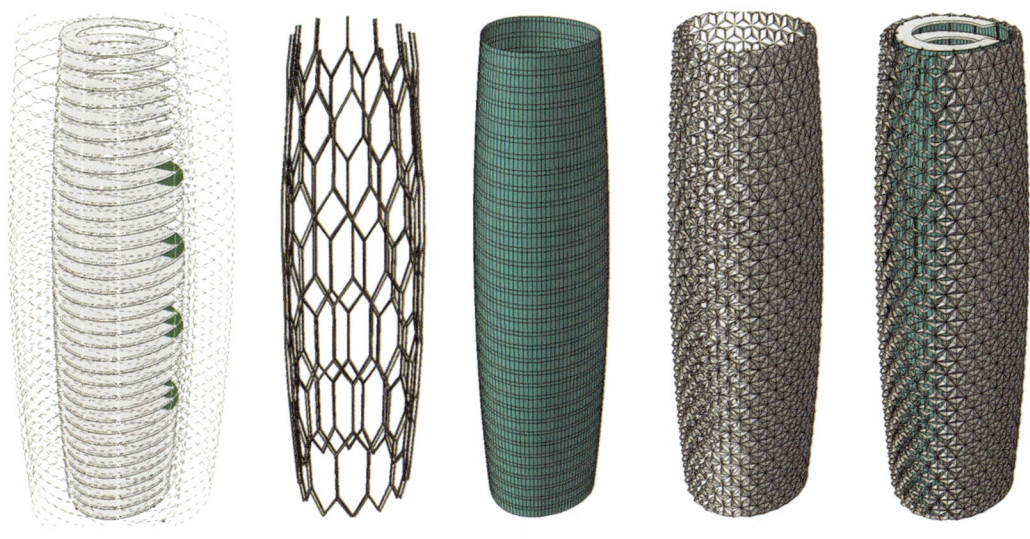

Figure 5.14 BIM model built in Digital Project in collaboration with Gehry Technologies.

Figure 5.15 Mashrabiya shading screen unit kinetic proof of concept parametric model built in Digital Project in collaboration with Gehry Technologies using circles and spheres to track the motion trajectory of the moving components.

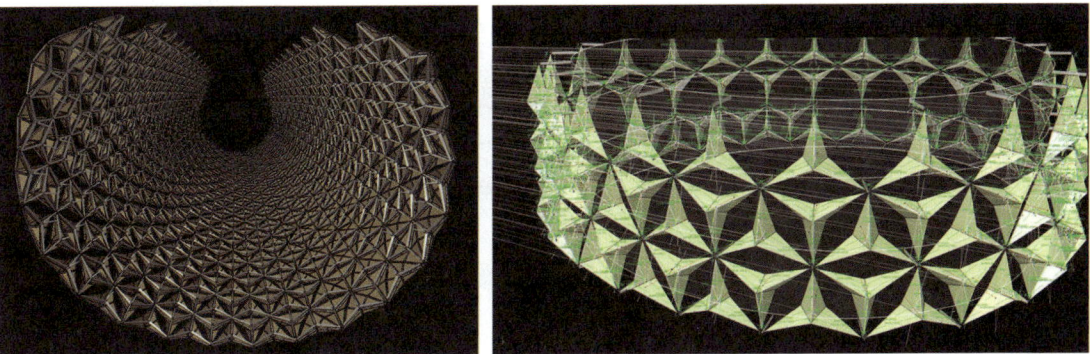

Figure 5.16 Left: Fully modeled and populated kinetic Mashrabiya units. Right: First solar rays study where the initial adaptive rules and assumptions were tested and analyzed.

most exciting and distinguishing thing about the project was to think about how kinematics details of a few millimeters could determine the performance of the entire façade, and then to calculate with functional precision how the actuators of the shading devices should behave."

With the first Mashrabiya unit kinetic model worked out, it was populated around the building in preparation for the first solar rays study (Figure 5.16). The results of the first study were not satisfactory. It showed extreme results where the units are either fully folded or fully unfolded. Intermediate positions were not utilized following our previous rules. It was therefore necessary to go back to the drawing board. In that respect, there was a need to optimize the geometry of the building, the shading units, and the geometric rules of the adaptive principles with respect to the solar rays. There was equal need to develop a communicable format to convey the new optimized design and associated principles across the project team.

Adaptive Principles Optimization, Construction, and Performance Manual

Figures 5.17 and 5.18 are self-explanatory slides extracted from the Al-Bahar Towers Geometry Construction & Performance Manual that

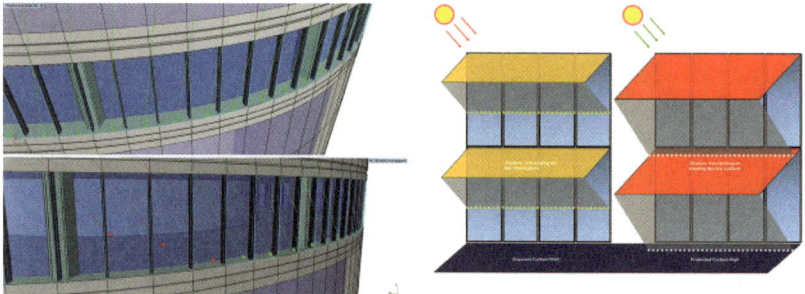

Figure 5.17   Left: This diagram shows that no direct solar rays (red) are allowed inside the occupied floor space. Right: A shading system works effectively when the shadow of one shading component casts all its shadow edges on the neighboring shading unit (as is the case on the right).

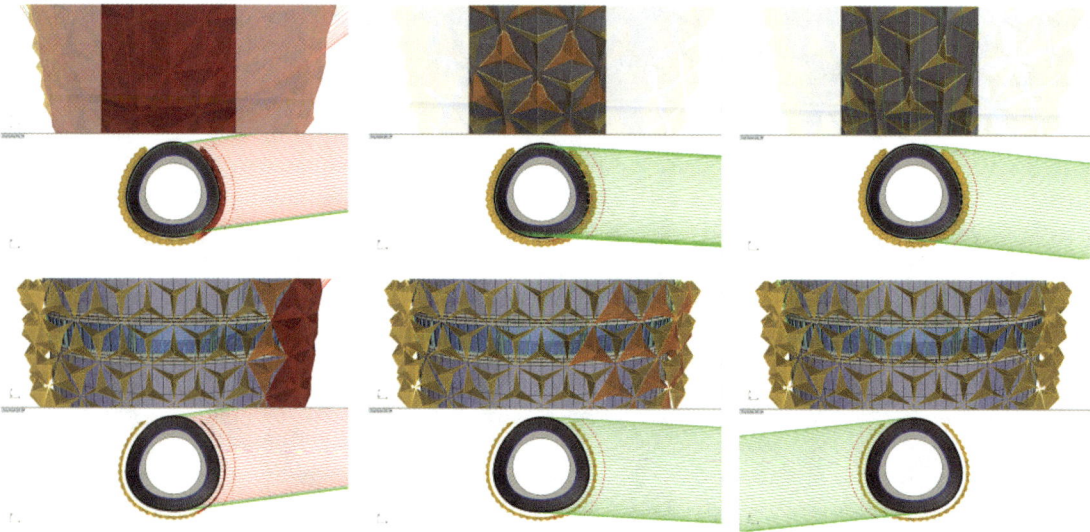

Figure 5.18   Top left: At low sun angles, the shading units unfold completely to block direct solar rays. Top middle: At intermediate solar angles, the units are deployed in mixed folding positions. Top right: At high solar angles, the units are deployed in maximum fold position to block direct solar rays. Bottom left: At low sun angles. Bottom middle: At intermediate sun angles. Bottom right: At high sun angles.

demonstrate the adaptive principles of the folding/unfolding of the Mashrabiya shading units with respect to the changing solar rays angle.

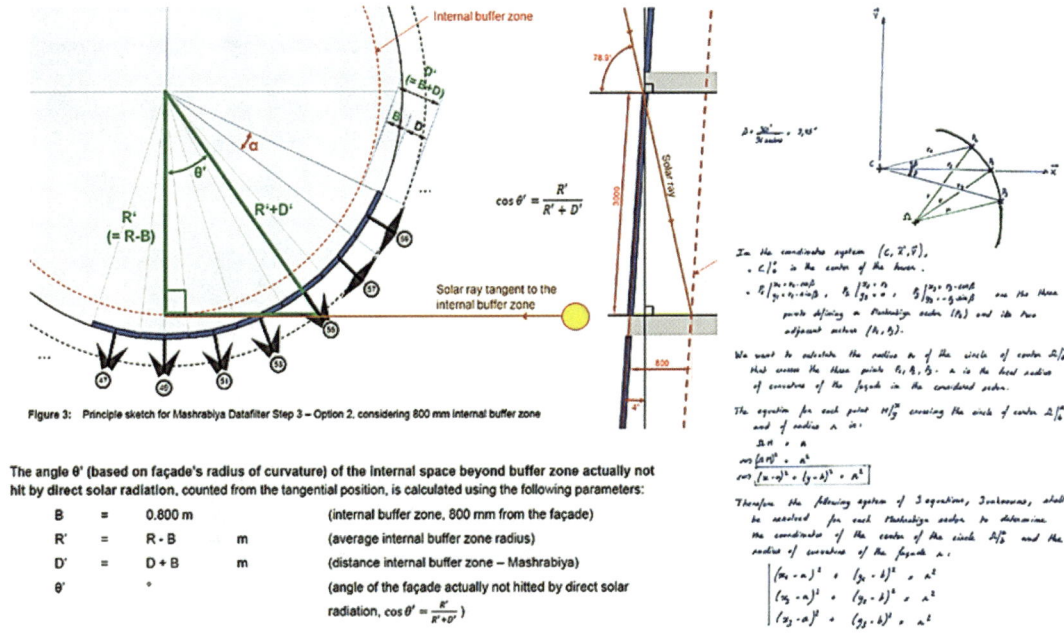

Figure 5.19   Principles used by Yuanda Europe, façade specialist engineers, to develop the code embedded in the Siemens HMI control software that runs the Mashrabiya shading system.

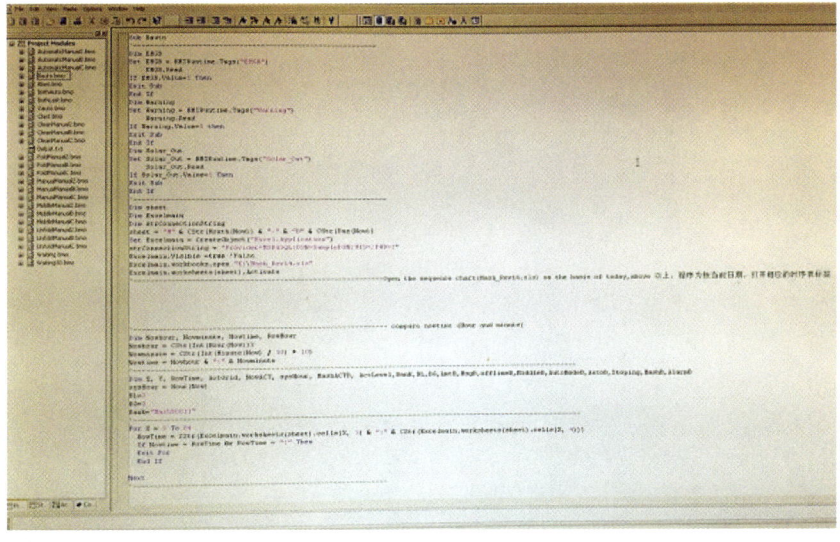

Figure 5.20   Part of the code embedded in the Siemens HMI software of the Mashrabiya shading system.

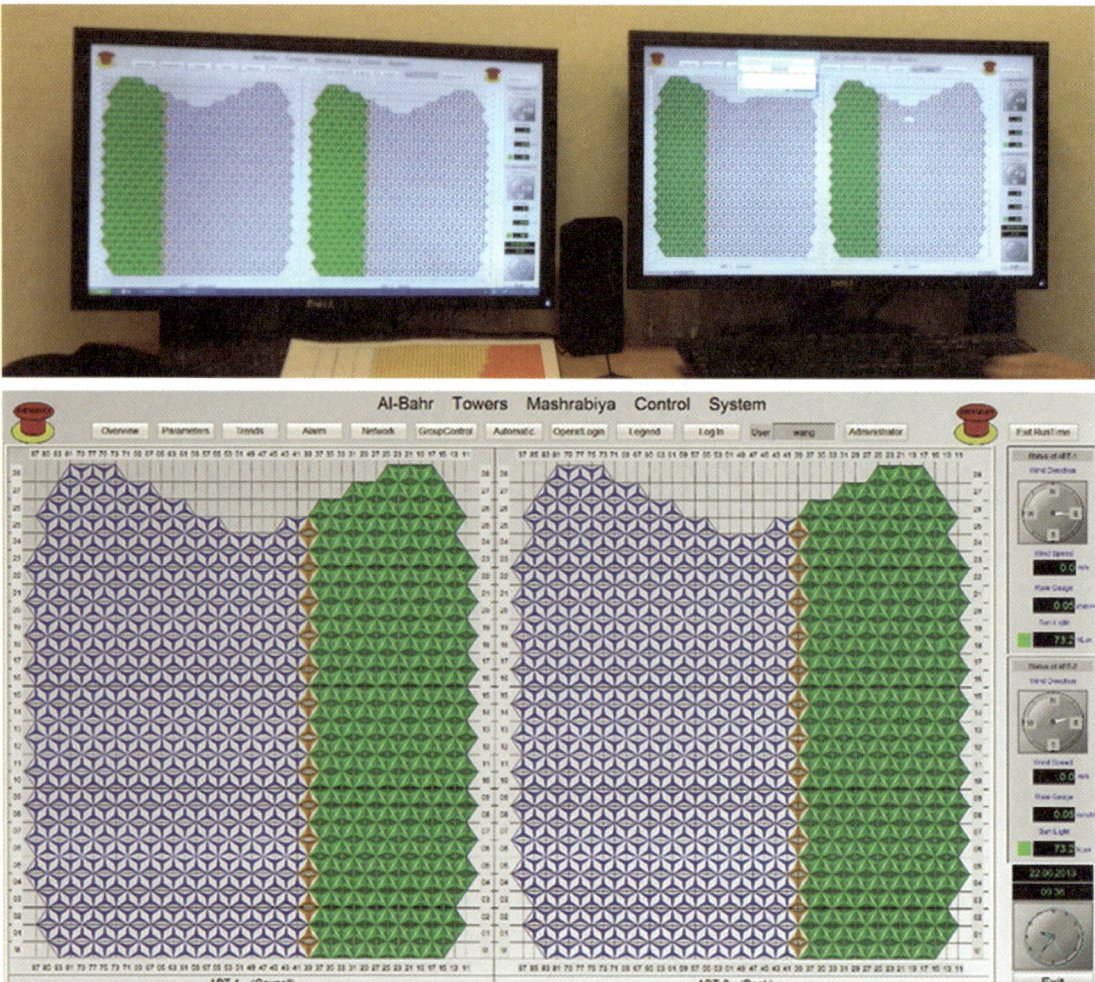

Figure 5.21   Top: BMS control room operator. Bottom: Mashrabiya HMI software close-up look.

Updating the Adaptive Algorithmic Principles and HMI Control Software

In summary, the Mashrabiya unit kinetic model (Figures 5.19–5.21) is self-explanatory in terms of the mathematical rules of the adaptive principles that drive the folding/unfolding of the Mashrabiya units.

Chapter 6

Custom-Designed Structures and Façades with Parametric-Algorithmic BIM Systems: 1 Bligh Street, Green Star Rated High Rise Project, Sydney

Thomas Spiegelhalter

Introduction

In 2006, ingenhoven architects from Germany, in association with Architectus from Australia, won the DEXUS/City of Sydney design excellence competition for 1 Bligh Street—a new AU$270 million, 135.5m high and 28-story elliptical office tower for the property groups DEXUS Property Group, DEXUS Wholesale Property Fund, and Cbus Property. The team's submission was chosen from 26 competitors from 17 countries. The realization planning and building process started in 2009 and included the collaboration of ingenhoven architects and Architectus, DS-Plan, Arup, Enstruct, Cundall, Grocon (builder), DEXUS Property Group, DEXUS Wholesale Property Fund and Cbus Property.

Since the project completion in 2011, 1 Bligh Street has won numerous awards such as the Council on Tall Buildings and Urban Habitat award for most outstanding new tall building in Asia and three awards from the Australian Institute of Architects for best commercial building, sustainable building, and urban design. Additionally, it was the first high-rise office tower in the southern hemisphere of this specific climate zone to feature double-skin glass façade technologies. DS-Plan, Arup, and Cundall assisted the design project team in pursuing these innovative integrated sustainable solutions to create an optimal indoor environment and overall

Figure 6.1 Multiple views from the street and sky level, interior and atrium levels, and views from the top terrace level to the bay area. (Image courtesy of ingenhoven architects, Düsseldorf, Germany, 2013.)

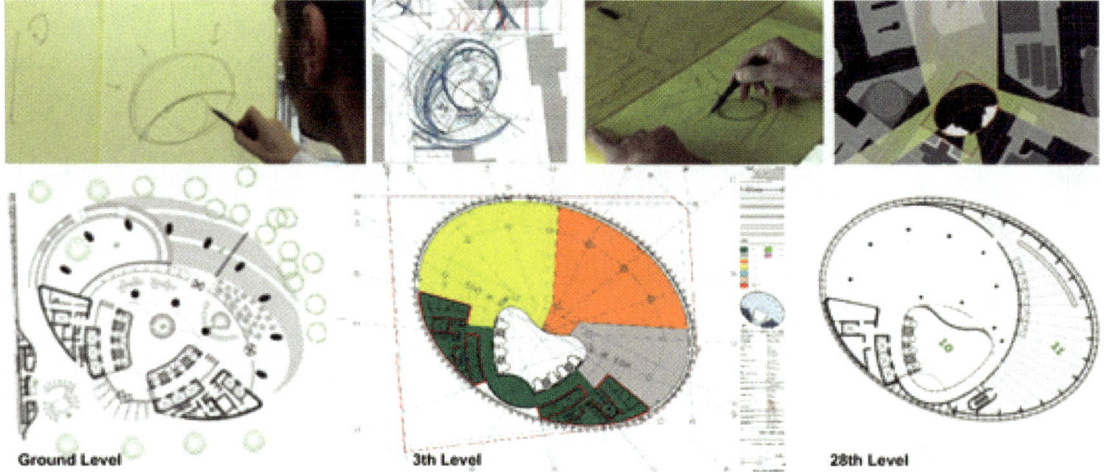

Figure 6.2 Conceptual site plan and building configuration sketches and plans by Ingenhoven and Architectus. (Image courtesy of ingenhoven architects and Architectus, 2008.)

efficiency rating for Australia's first-awarded 6 Star Green Star high-rise building and a 5 Star NABERS energy rating. The building also won the prestigious International High-Rise Award 2012 in Germany (Figure 6.1) [1].

Site Condition, Building Key Features, and Systems

The 1 Bligh Street high-rise office tower has a gross area of 45,760

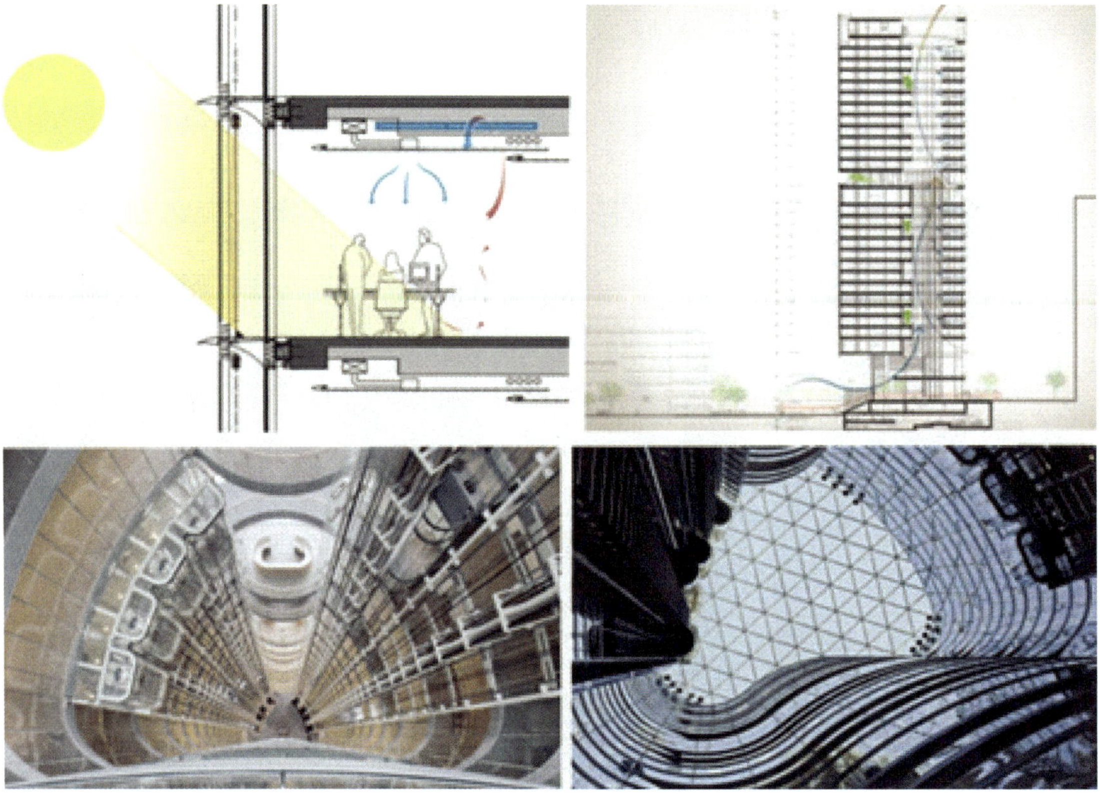

Figure 6.3 Naturally ventilated atrium and façade section. (Images courtesy of ingenhoven architects and Architectus.)

m² and is located in a prominent location in the heart of Sydney. The site context and its potentially wide range of far-reaching views were an important design factor for ingenhoven and Architectus determining the elliptical shape of the building and its orientation (Figure 6.1). Its elliptical geometry elegantly resolves the site's urban condition at the confluence of two city grids, as documented in the conceptual sketches of ingenhoven and Architectus in Figure 6.2. The ellipse is rotated on its site to address views to—and, importantly, from—the harbor, simultaneously resolving the urban condition and maximizing the building's commercial value. Due to its compact geometry and a slight rotation of the building in relation to its site, all offices enjoy unobstructed views of the city's beautiful harbor. All the typical functions of the ground floor have been raised to allow a public plaza at street level and allow public movement through and around the site.

Most of the spatially designed 28 levels contain 1,600 m² large floor plates with column-free uninterrupted office space enjoying views of Sydney Harbor. Bathrooms and fire stairs enjoy views across the city skyline. The floor plates are interconnected by stairs, bridges, and balconies. Levels 15 and 28 contain external terraces with native planting areas.

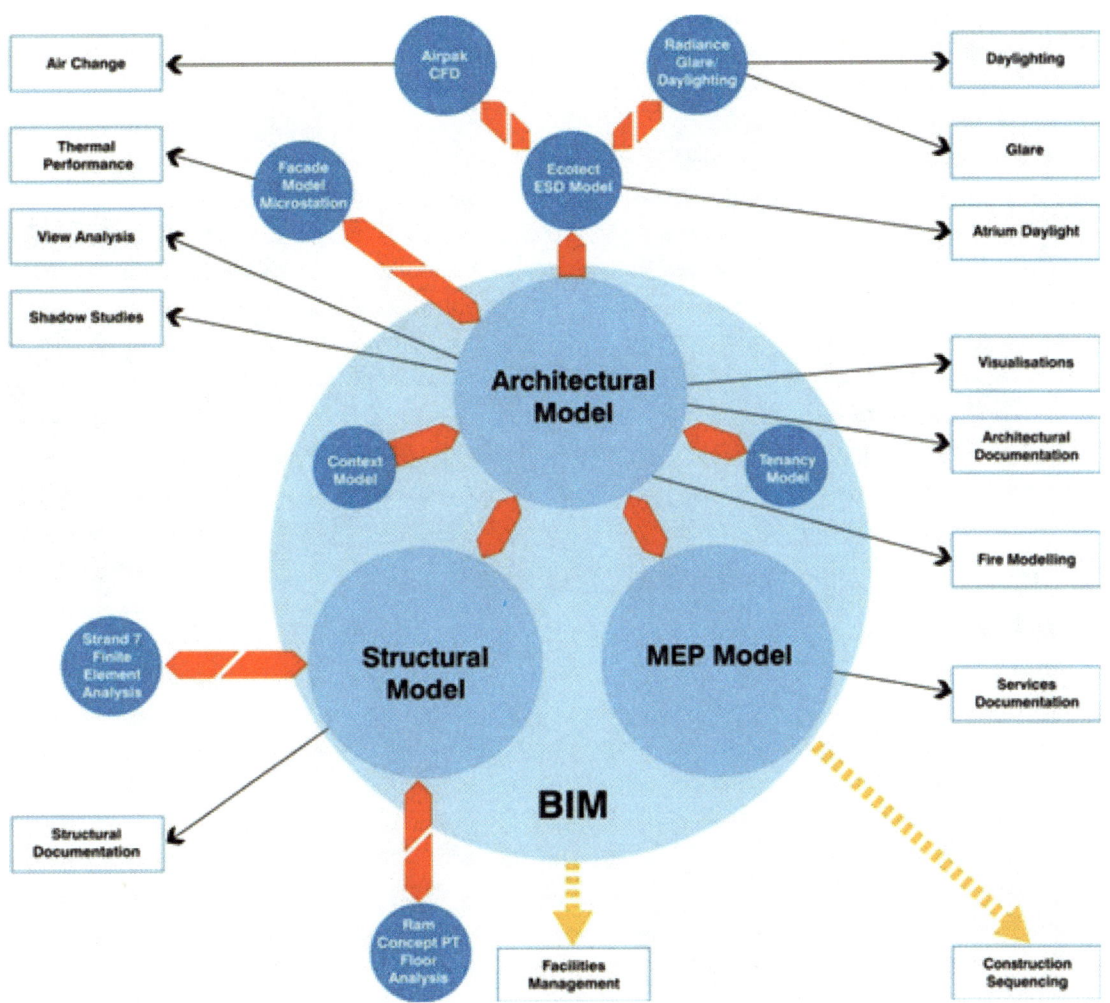

Figure 6.4 BIM conceptual flow diagram for 1 Bligh Street Project. (Image courtesy of Architectus and ingenhoven architects.)

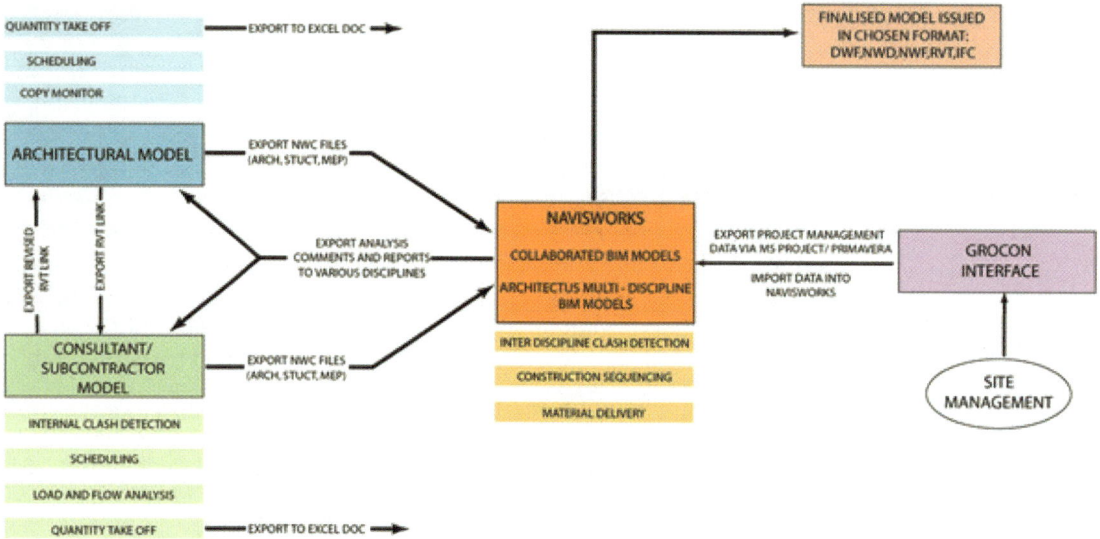

Figure 6.5 BIM management flow diagram for 1 Bligh Street Project. (Image courtesy of Architectus, 2013.)

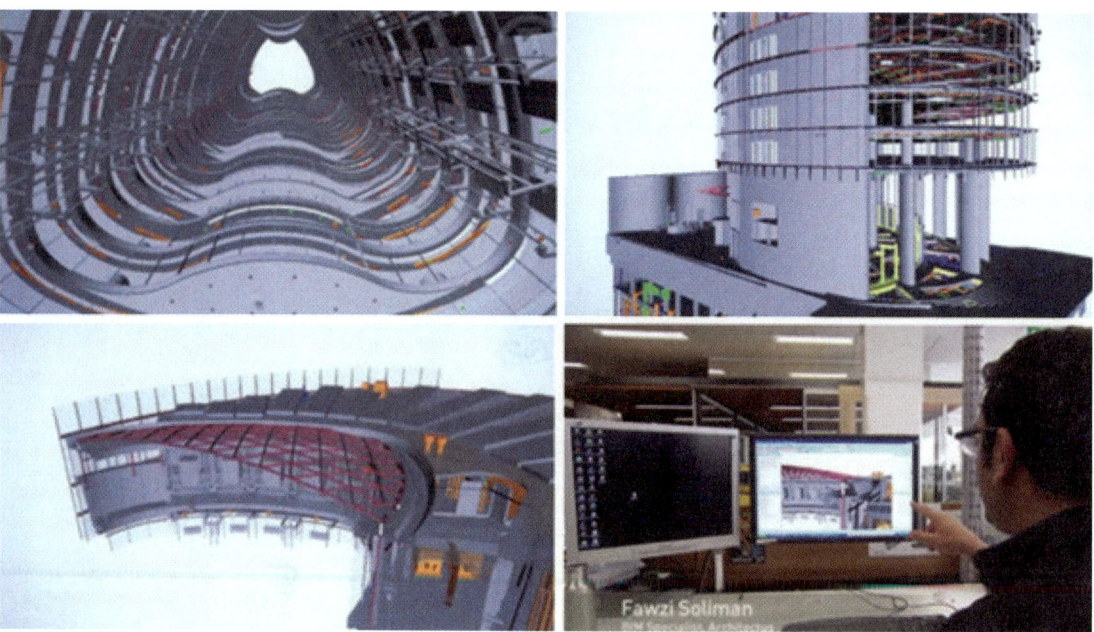

Figure 6.6 BIM process images by Architectus, Sydney. (Image courtesy of Architectus and ingenoven architects, Source: http://www.archi-ninja.com/the-making-of-1-bligh-street-sydney/.)

The 'heart of the glass tower' is a naturally ventilated atrium that extends from the ground floor to the rooftop. This is reinforced by the full-height fresh-air atrium that sits at the core of the building. By drawing air in at the bottom of the atrium and through the glass louvers of the south façade, the atrium acts as a giant cooling chimney. It also offers that rarest of commodities in an office building: genuine fresh air for the workers, since each floor opens to this space as an internal balcony (Figure 6.3).

Double Façade, Space Conditioning, Cooling, and Energy Use Concept

One of the most innovative features included the first use of an automated double-skin glass façade of this scale in Australia, a naturally ventilated 28-story atrium with highly efficient, automated sun control and mechanical cooling systems for Sydney's hot, sunny environment. The highly effective façade system automatically adapts through the integrated building automation system to ever-changing solar conditions at each orientation to minimize the solar penetration to the floor and control direct glare. This allowed the use of very clear glass, creating a building that is very transparent from the outside looking in and from the inside looking out (Figures 6.1 and 6.3).

The ventilated cavity contains automatically controlled, view-preserving louvers that mean the sun never hits the inner skin, so that thermoactive, chilled-beam radiant cooling provides the building occupants with dynamically changing indoor conditions along the perimeter zone for the required thermal comfort settings.

Challenges of the Multidisciplinary CAD to BIM Collaboration

1 Bligh Street was one of the first commercial projects in Australia to implement multidisciplinary BIM collaboration between architects, engineers, contractors, manufacturers, and clients and was set from the very beginning of the planning process as a contracted requirement. In this context, in 2008 Architectus won Autodesk's International Revit BIM Experience Award for its successful multidisciplinary team coordination that integrated structural design analysis, mechanical, electrical, and other engineering files, and used built-in-synchronization and an interference checking tool to reduce errors (Figures 6.4 and 6.5). In 2013 Architectus won the prestigious American Institute of Architects COAA Owners' Choice BIM Award for New Construction for its work on the project.

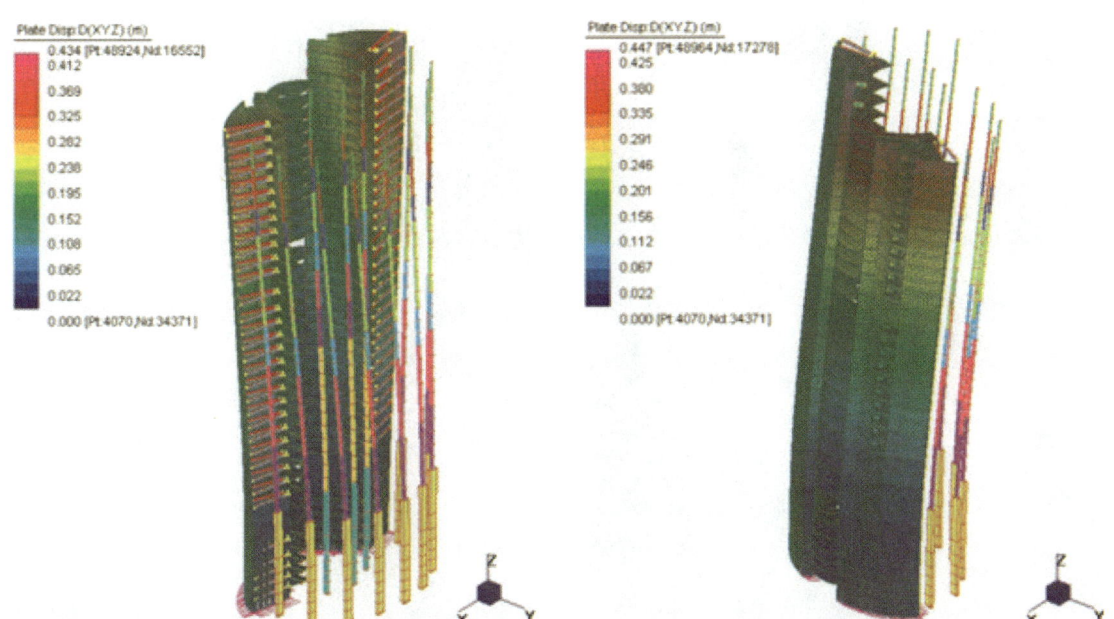

Figure 6.7 Finite element analysis of the structural design using Strand 7. (Image courtesy of Tim Boulton, Enstruct, Australia.)

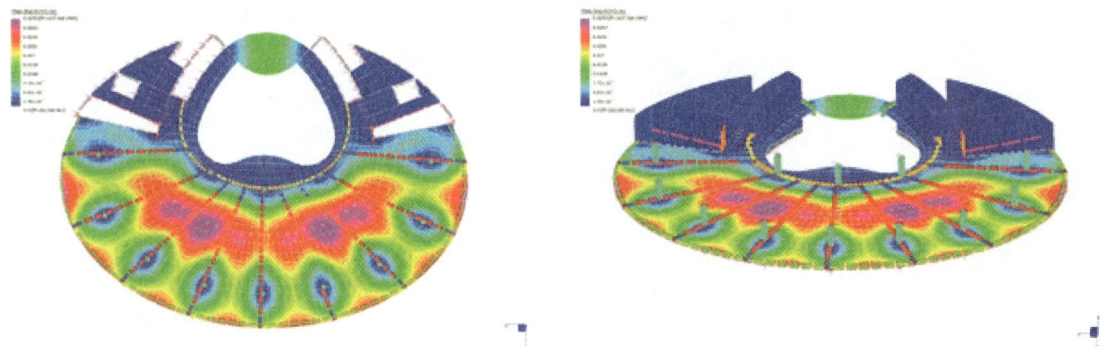

Figure 6.8 Analysis and design of post-tensioned and reinforced concrete floor systems using RAM Concept. (Image courtesy of Tim Boulton, Enstruct, Australia.)

The collaboration between ingenhoven architects in Düsseldorf, Germany, and Architectus in Sydney, required an efficient data exchange management for their differences between their design file use and documentation methodologies. While Architectus was using BIM procedures, Ingenhoven did work on a 3-D basis but did not use BIM at the time of the planning stage. Ingenhoven architects' design work on 1 Bligh Street was produced with Bentley Microstation® and forwarded to Architectus who ran the 3-D-BIM model in Autodesk Revit® Architecture. This model was then supplied to Enstruct as the starting point for its model in Autodesk Revit® Structure 2008. Importing and exporting files between the disciplines was seamless in native Revit format and thus did not require the use of Industry Foundation Classes (IFCs), although this was available as an archive format. The 3-D-model visualization assisted the multidisciplinary

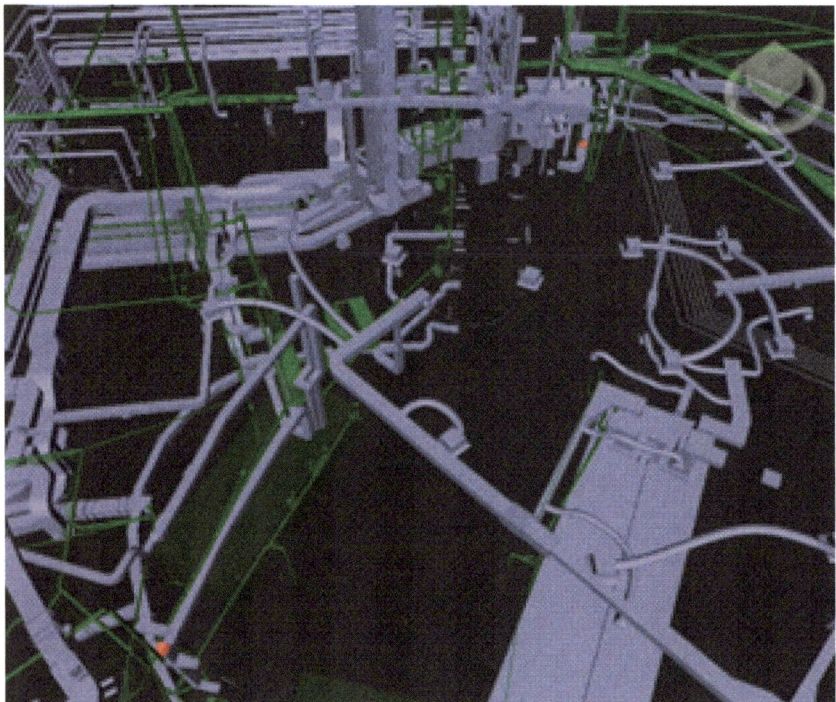

Figure 6.9 Mechanical services of a typical floor using Revit Services. (Image courtesy of Architectus.)

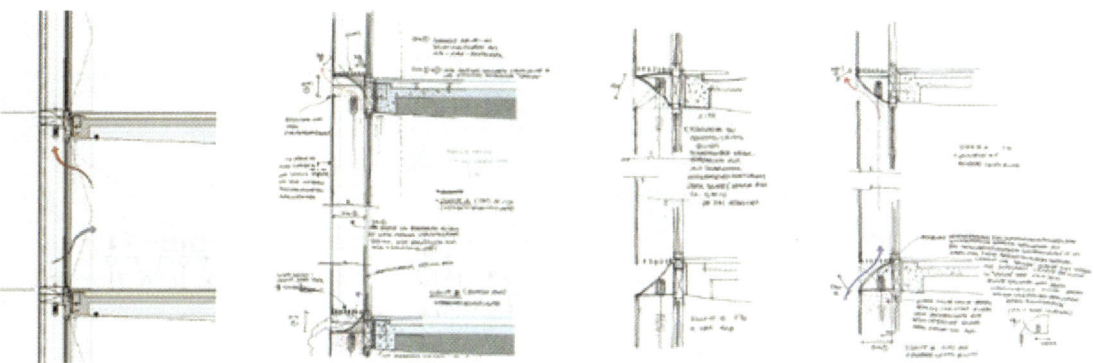

Figure 6.10 CFD simulations preliminary results with shaped aluminum louvers and impacts on technical services (possible heat buildup, cooling loads, etc.). (Image courtesy of DS-Plan, ingenhoven architects, and Architectus, 2013.)

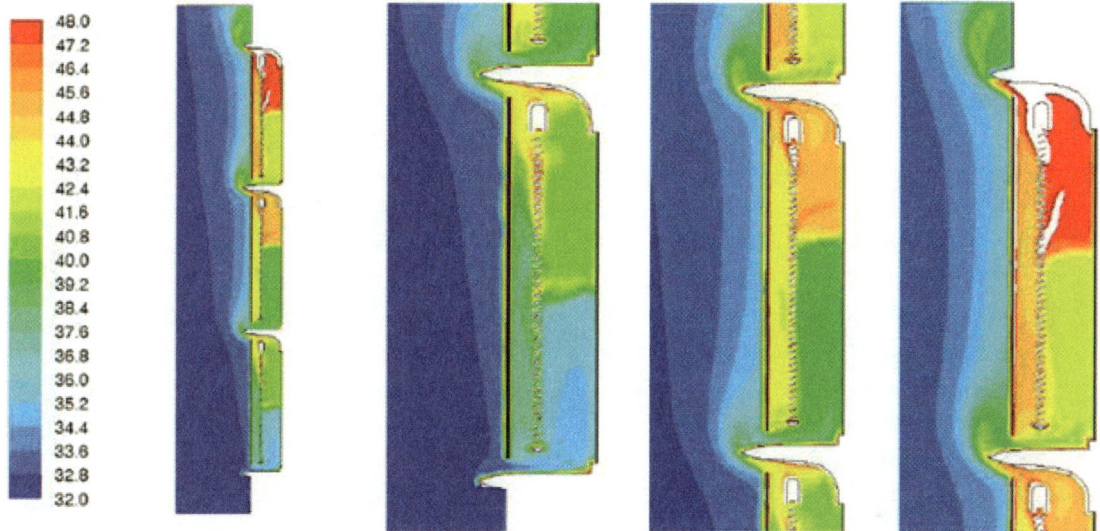

Figure 6.11 CFD simulation with preliminary results with open ventilation openings. (Image courtesy of DS-Plan, ingenhoven architects, and Architectus, 2013.)

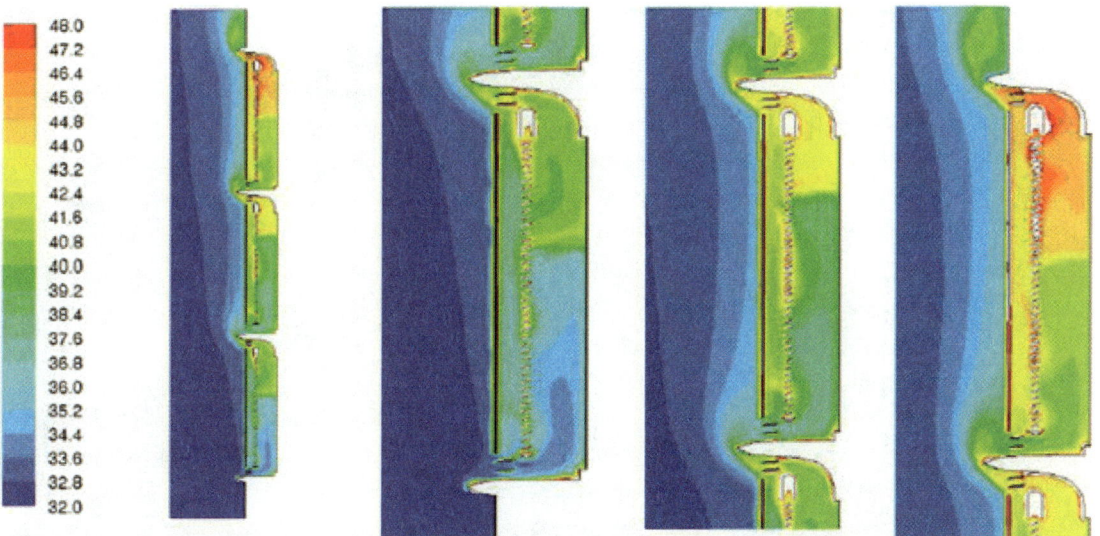

Figure 6.12 CFD simulations preliminary results with shaped aluminum louvers and impacts on technical services (possible heat buildup, cooling loads, etc.). (Image courtesy of DS-Plan, ingenhoven architects, and Architectus, 2013.)

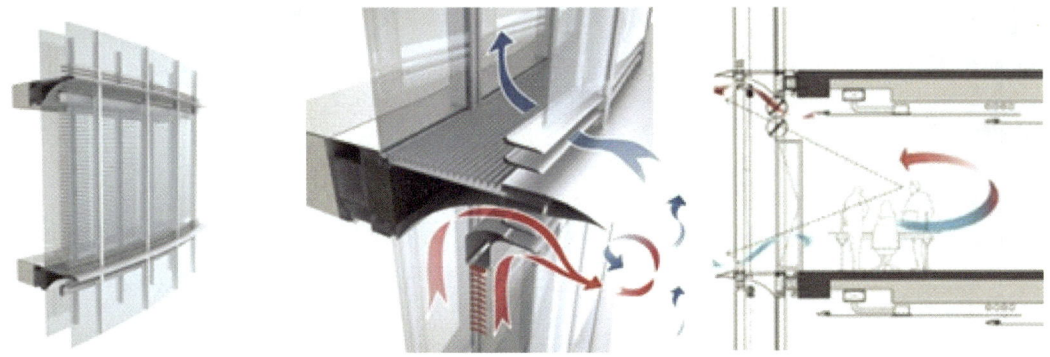

Figure 6.13   Façade details and renderings by ingenhoven architects. (Image courtesy of ingenhoven architects, 2013.)

collaboration, in particular for the 4-D floor space area (FSA) analysis and schedules, which could not be achieved by a standard CAD system.

During the construction stage, Architectus was appointed as BIM managers to the project, creating a BIM execution plan for it and combining the model inputs in a number of 3-D and BIM formats from all trades and disciplines into a Navisworks model which, updated regularly, formed the basis of communication and clash detection for the project. BIM reduced the number of requests for information (RFIs) through a better understanding and visualization of the building. More than 10,000 identified clashes were dealt with during the construction period. Where subcontractors did not have sufficient experience with BIM to create their own models, Architectus seconded staff to them to both train and assist them with their workload. In some cases Architectus modeled on behalf of subcontractors.

Figure 6.14   Façade mock-up by G. James, Australia. (Image courtesy of ingenhoven architects, 2013.)

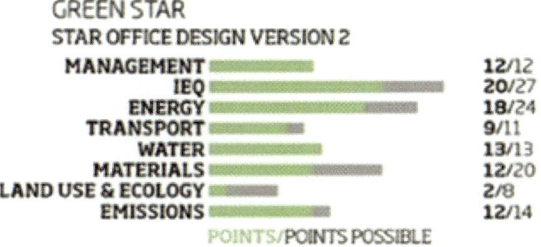

Figure 6.15   The Australian Green Star Rating for 1 Bligh Street. (Image courtesy of ingenhoven architects and Architectus, 2013.)

The handover model provided to the client in Navisworks format contains 35 individual models representing all 32 disciplines.

Structural Analysis and Design Integration

The involvement of Enstruct Group Pty Ltd, NSW Australia, as structural engineers started at the Stage 1 development application (DA) and continued through design competition, lodging of Stage 2 DA, and preparation of the tender documentation. Enstruct was then engaged by the main contractor Grocon to take the design through to construction and completion. Enstruct was responsible for the entire superstructure and foundation floor plates, and columns. The 3-D visualization in the BIM-model allowed Enstruct to better develop structures in the overall context,

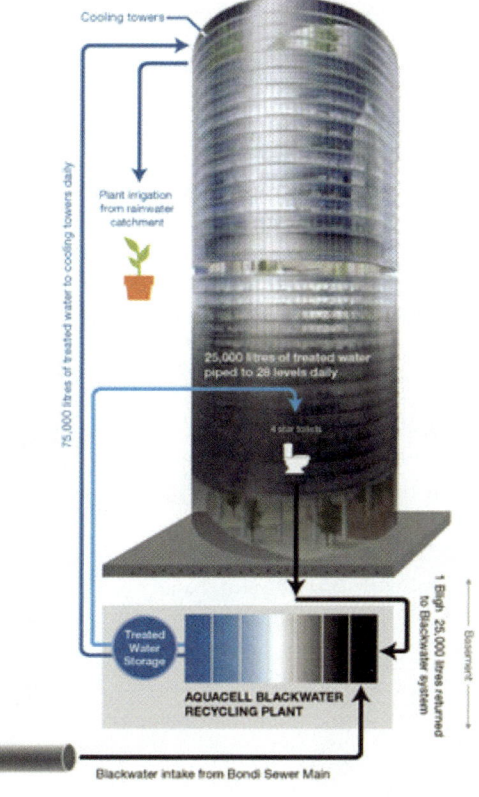

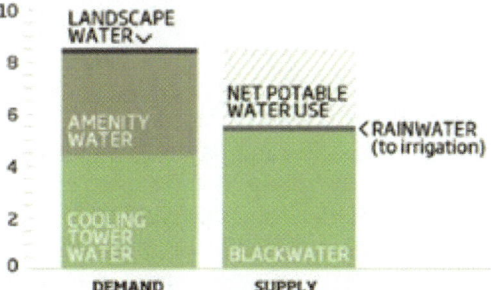

Figure 6.16   Diagram of the central black water treatment plant and annual net water use of 1 Bligh Street. (Image courtesy of DEXUS Property Group, ingenhoven architects, and Architectus, 2013. Source: http://dexus-crs10.reportonline.com.au/envproperty.php.)

as noted by the director Tim Boulton: "Being able to look at something in 3-D gives you a far better understanding of what is going on, rather than an old-fashioned 2-D plan" [4]. Enstruct used for the specific structural optimization process for the lateral analysis of the building Strand7®, and for the floor plan analysis used Bentley Ram Concept®18. For the beam element analysis and optimization, the engineers used a system called Space Gass with a link between Space Gass and Revit® to increase the interoperability. The combination of these software tools enabled the highest optimization and scenario outcomes for the automated process of multiple layouts of design spans, and automated generation of critical section considering actual—not simplified—geometries for automatically generated load combinations for each required design code (Figures 6.10 and 6.11).

However, according to Tim Boulton of Enstruct the export and import of the structural analysis files into Revit® were often troublesome and content needed to be recreated. Thus, the architects continuously preferred the analysis to be made in a separate model, with any changes first reviewed and approved by the structural engineers and then translated into the collaborative 3-D BIM model. Monitoring tools were used to track and coordinate changes between the structural and architectural models [4].

Integrated Double Façade Performance Analysis, Mechanical, Electrical, Plumbing, and Fire Service Design

Arup in Sydney, Australia, was responsible for the mechanical, electrical, and fire services, as well as being the façade design consultants with DS Plan from Stuttgart, Germany. While the engineers used Revit® MEP Services to model a typical floor, it was not used to produce the documentation. Service documentation was done in AutoCAD® and AutoCAD® MEP®. As recognized by the services engineers and the rest of the 1 Bligh Street team, mechanical, electrical, and plumbing (MEP) services were not part of the BIM model. Arup's project director Kerryn Coker identified the tight program and the complexities of producing services documentation using BIM as the two main reasons that prevented MEP services from being included in the BIM model. As opposed to the view of Architectus, the services engineers felt that the out-of-the-box content included was poor, and developing it would take a considerable amount of time with an already tight deadline. Nevertheless, Arup did isolated simulations, including a thermal simulation, to calculate the sizing of the air conditioning equipment [5].

DS-Plan, Stuttgart, Germany, conducted parallel numerous computational fluid dynamic (CFD) simulations of the thermal performance behavior of the integrated sun, glare and natural ventilation control systems of the double façade and superstructure components in collaboration with ingenhoven architects, who developed all the preliminary and overall planning details for the design-to-production processes (Figures 6.9–6.12).

Interoperability with the Contractor and Subcontractors

As previously mentioned, BIM was also employed by the general builder Grocon to manage the documentation of the base building and the integrated fitout. Coordination sessions were conducted with consultants and subcontractors at which the combined 3-D/4-D services Autodesk

Navisworks model was used to select views, including conducting fly-throughs, in order to optimize the planning, bidding, manufacturing, and implementation process (Figure 6.8). The interoperability of the multidimensional BIM model allowed walking around in the 3-D model with design directors, clients, and consultants and engaged a far more incisive conversation than a 2-D paper output could ever provide [1].

The BIM manager was aware that subcontractors had to remodel the Revit® model provided by the consultants because it was not compatible with the subcontractors' specific manufacturing software. Allan Hickey, general manager of the Double Glass Façade Builder G. James in Brisbane, Australia, described such challenges in the differences of the software use: "In general, the delivered Bentley, Autodesk AutoCAD and BIM files were not helpful for our production. We had to redraw all the drawings with Autodesk AutoCAD and for production we used Autosketch. The architects delivered a pure ellipse, each panel was unique. We needed to rationalize to a series of circular arcs. Then the advantage was that all the panels could be mirrored and automatically repeated in the design-to-fabrication processes. The automated repetitions with Autodesk Sketch for our shop drawings helped to feed the NC machines. Today, we use a paperless production and factory process with Autodesk SP Inventor 3-D models" [6].

Renewable Energy, Water Recycling, and Benchmarking

1 Bligh Street uses a natural gas-fired trigeneration system to simultaneously produce three forms of energy: electricity, heating, and cooling. This reduces energy loss through transport and provides greater operational flexibility for the operation for the building with demands for heating as well as cooling. The trigeneration systems also increase the energy efficiency by capturing waste heat through heat-energy-recovery systems that result from the power generation that can be then used to generate hot and cold water. In addition, solar energy tubes system located on the roof provides thermal energy for hot water, and assist an absorption chiller for radiant cooling throughout the building.

The overall simulated annual purchased energy use is predicted by 105, 8 kWh/m² and the annual carbon footprint is 30 kg/CO2e/m². The Australian Green Star Rating for the Office Design Version 2 arrived with 18/24 points for Energy, 13/13 points for water and 12/14 points for emissions rating (Figure 6.14).

According to DEXUS Property Group/DEXUS Wholesale Property Fund/Cbus Property (building owner), 1 Bligh Street treats wastewater through the first private retailer's license in NSW for a central blackwater treatment plant. The technology prevents wastewater from being pumped into the ocean. Blackwater recycling reduces water consumption via the provision of clean recycled water for the washroom flushing system. This means about 100,000 liters of water are saved per day, the equivalent of one Olympic swimming pool every two weeks. Recycled rainwater is used to irrigate a variety of plantings spread throughout the building. Water-efficient appliances are used throughout, with 90% of the water demand met by recycled water.

Conclusion

One of the major challenges in transferring parametrically developed geometries to the 3-D BIM and 4-D Naviswork model was to define its level of detail. Any exchanged big file size made navigating and applying changes to the model a slow process and sometimes even hindered it. Architectus mostly detailed typical spatial, geometrical complexities and floors and relied the model on 2D line work for detailing anything over 1:20 scale drawings, whereas ingenhoven architects continuously and parallel developed detailed 2-D plans in the scales of 1:10, 1:5, and 1:1 for the contractors and manufacturers to the very end of the project. 3-D BIM or 4-D Naviswork did not include details below 1:20 scales for contractors to read, for example, important façade gasket and screws scales and properties for their manufacturing process.

Conversely, this approach contradicts the generally claimed principle of BIM as a single-model database. On the other hand the hybrid approach allowed producing different 3-D models with different levels of 2-D and 3-D detailing depending on their use.

However, the hybrid advantage for repetition and automation of designs, details, and sections resulted from the fact that the architects delivered a pure ellipse so that all the building systems and façade components could be mirrored and automatically repeated in the design-to-fabrication processes. The combination of parametric and compatible software tools of the engineers enabled the highest optimization and scenario outcomes for the automated process of multiple layouts of design scenarios, automated generation of critical section considering actual—not simplified—geometries for automatically generated combinations for each required design load and code.

In summary, the project demonstrated that even with an earlier generation of 3-D BIM software it was possible to gain many benefits from this hybrid approach of different software tools to feed BIM at the multiple design development stage of the project. While there were numerous limitations to software and workarounds required, solutions were found to allow the optimization of the design from the shared information provided by BIM-based simulation and analysis. The value of this combination is exemplified in managing geometric complexity, developing subcontractor designs, resolving clashes to prevent rework management of the construction schedule, and as a communication tool that resulted in a substantial drop in requests for information and the near elimination of contract sum adjustments. In addition, the adoption of BIM by a wide group of subcontractors and the head contractor has had a permanent effect in uplifting the skills base of the local construction industry.

Participants
Client: DEXUS Property Group, DEXUS Wholesale Property Fund, Cbus Property, Australia
Team ingenhoven architects, Düsseldorf, Germany
Design Team: Christoph Ingenhoven, Martin Reuter, Christian Kawe, Hinrich Schumacher, Martin Slawik, Thomas Weber, André Barton, Mario Böttger, Elisabeth Broermann, Darko Cvetuljski, Ralf Dorsch-Rüter, Hye Jin Jung, Christian Kob, Andrea König, Alice Koschitzki, Peter Pistorius, Dr. Mario Reale, Evelyn Scharrenbroich, Ulrike Schmälter, Alexander Schmitz, Jürgen Schreyer, Brett Stover, Erich Tomasella, Lutz Büsing, Felix Winter

Team Architectus, Sydney, Australia:
Design Team: Ray Brown, Mark Curzon, Simon Zou, Linda Bennett, Scott Hunter, Ryan Townsend, Daniela Salhani, Nikhil Fegade, Karolin Baer, Tommy Ford, Fawzi Soliman, Michael Harrison, Stewart Verity, Murray Donaldson, Camille Lattouf, Rodd Perey, Ryan Hanlen, Harry Broekhus, David Kamel, Darrin Rodrigues, Chase Ronge, Siera Chuah, Annette Gall
Structural Engineering: Enstruct Group
Façade Consultant: DS-Plan, Arup, and Enstruct,
Ecologically Sustainability Design: Cundall, Sydney, Australia
Mechanical Services, Fire Protection-Dry Fire, Acoustic and Electrical Services: Arup
Lighting Artificial: Arup, Tropp Lighting Design, Germany

References

[1] The Making of 1 Bligh Street, Sydney. http://www.archi-ninja.com/the-making-of-1-bligh-street-sydney/#sthash.dwCeve4G.dpuf, retrieved on Nov. 25, 2013.

[2] Charles Linn. Down Under, A Highly Sustainable High-Rise, ASME. https://www.asme.org/engineering-topics/energy-efficiency/down-under-a-highly-sustainable-high-rise, retrieved on Nov. 25, 2013.

[3] The Making of 1 Bligh Street, Sydney. http://www.archi-ninja.com/the-making-of-1-bligh-street-sydney/.

[4] Interview with director Tim Boulton of Enstruct Group Pty Ltd, on Nov. 28, 2013.

[5] Interview with Kerryn Coker, Ove Arup, Sydney, on Dec. 4, 2013.

[6] Interview with Allan Hickey, Brisbane, on Nov. 27, 2013.

[7] Dexus: Energy Initiatives Report. Source: http://dexus-crs10.reportonline.com.au/envproperty.php, retrieved on Nov. 25, 2013.

[8] Grocon, Naviswork Model for 1 Bligh Street, http://usa.autodesk.com/adsk/servlet/item?linkID=14271589&id=11681286&siteID=123112, retrieved on Nov. 25, 2013.

Chapter 7

Parametric-Algorithmic Automated Modeling and Fabrication: The Railway Station Stuttgart 21

Albert Schuster, Lucio Blandini, and Thomas Spiegelhalter

Urban Large-Scale Project Main Station Stuttgart 21

The Stuttgart 21 project of the award-winning office of ingenhoven architects is one of Germany's and Europe's largest urban renewal projects. As with other transformative large-scale urban projects before in history, this one also triggered many residents to be at first deeply opposed toward the multibillion euro undertaking. It will see the train tracks placed underground cutting through the center of Stuttgart, creating entire new interconnected neighborhoods between the old and new parts of the city. Parts of the current railway station and the new underground station will be the nucleus of the new urban development of Stuttgart 21. The Schlossgarten (palace gardens) is the green heart of Stuttgart and will only gain importance as the city expands. Some of the existing rail yards

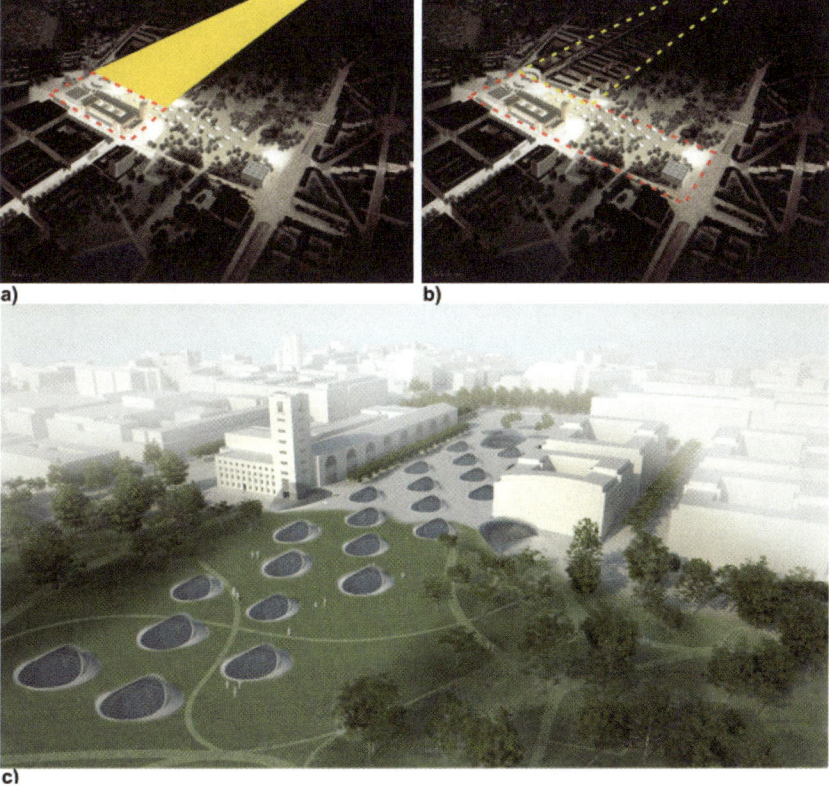

Figure 7.1 The old central station, a registered national landmark by architect Bonatz, will not be completely torn down. Important parts of it will be integrated into the new building cross section of the tracks. The project will turn the current above the ground terminus station into a subterranean through station. a) Design view on the existing terminal station and trackage highlighted in a yellow rectangle. b) New platform hall and city development on the trackage. c) Scheme of the new platform hall. (Source: Peter Wells and ingenhoven architects, 2013.)

Figure 7.2   Renderings of the new station. Top image: View into the platform hall. Images below from the left to the right: Cutaway model view, light eye in the park, cutaway model with a grid shell, 3-D animation of the backside of a light eye. (Source: ingenhoven architects, Düsseldorf, 2013.)

and remediated brownfields will become a new public park. The historical station building will function as a loggia, with its tower for the new railway station continuing to be a major landmark. This new zero-fossil-energy operated underground central train station will integrate of Stuttgart into the growing network of high-speed rail traffic in Europe (Figure 7.1). It follows ecological, carbon-neutral, socio-economic, and technological parameters and provides comfort, security, and job growth.

The following chapter will first focus on the integrated project delivery process of the parametrically designed and algorithmically scripted and automated double curved surfaces, modular shells, and lighting cones for the new railway station Stuttgart 21. The second focus describes the intelligent, adaptable zero-fossil-energy and CO_2 reduction design with automated control systems for comfort and security [1].

Parametric-Algorithmic Design of the New Railway Station Stuttgart 21

The present terminal station in Stuttgart will be replaced by a 420m long and 80m wide underground station designed by ingenhoven architects with consultant Frei Otto, the pioneer of the 1972 Munich Olympic Stadium. The new space is characterized by 28 double-curved concrete columns, the so-called chalices. The railway station was architecturally modeled with Bentley Microstation tools. The top surface of the station has five different slopes and the platform level has two different slopes in the longitudinal axis. These boundary conditions require different heights of nearly each

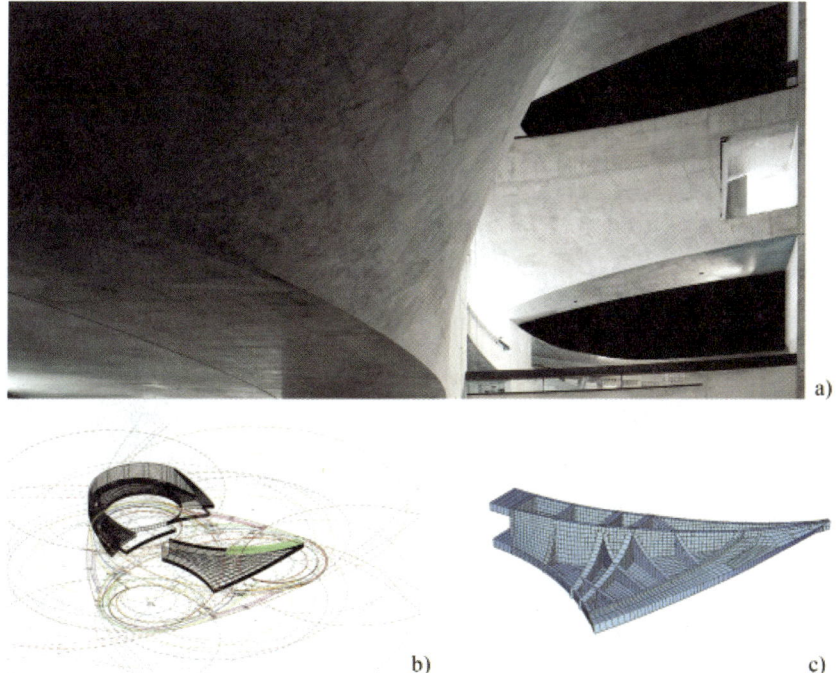

Figure 7.3 a) Mercedes-Benz Museum. View from inside (Brigida Gonzalez), b) Geometrical analysis (UN studio and Arno Walz, DesignToProduction), and c) Twist structural model. (Source: Werner Sobek, Stuttgart.)

chalice with a maximum amount of geometrically common elements in the concrete double-curved roof. The overall design of the entire railway structure enables a light and elegant concrete construction (Figure 7.2).

An interdisciplinary team at Werner Sobek's office modeled and calculated the most efficient buildable structure. In cooperation with the architects, the engineers set low-energy resourceful scenarios and defined assembly strategies [3]. According to Werner Sobek, there were several helpful precedents beneficial for the Stuttgart 21 railway station from their built project oeuvre for structural modeling and scripting. One of them is the new Mercedes-Benz Museum, Stuttgart, designed by UN Studio from Amsterdam, containing the most complex concrete elements [2]. The Mercedes Museum's ramp and the twist were engineered as curved hollow box girders with varying sections (Figure 7.3). In order to determine the concrete sections, the middle surfaces of the shells were derived based on the original outer surfaces, which were imported from an architectural Rhino model. The middle surface model was then meshed, analyzed, and structurally optimized with the finite elements software Sofistik. The thickness of the concrete walls was kept constant to 50 cm. In this precedent the formwork for the double-curved surfaces was prefabricated and costs were again minimized by the reuse of repeating elements. To maintain the correspondence between the geometric information in the 3-D model and the built surfaces, a cloud of reference points was defined to allow on-site verification of the respective coordinates [2].

However, the Stuttgart 21 project was in many ways more challenging because of the constantly varying concrete shell thickness. Werner Sobek's team had to identify and develop new methods and tools in order to precisely define the thicknesses at each node of the finite element mesh. Moreover, the radii of curvature were much smaller here compared

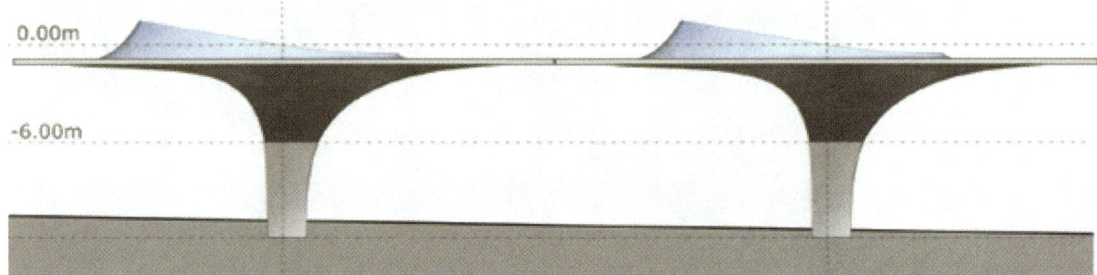

Figure 7.4 Section showing the division line for the chalices. (Source: Werner Sobek, Stuttgart, 2013.)

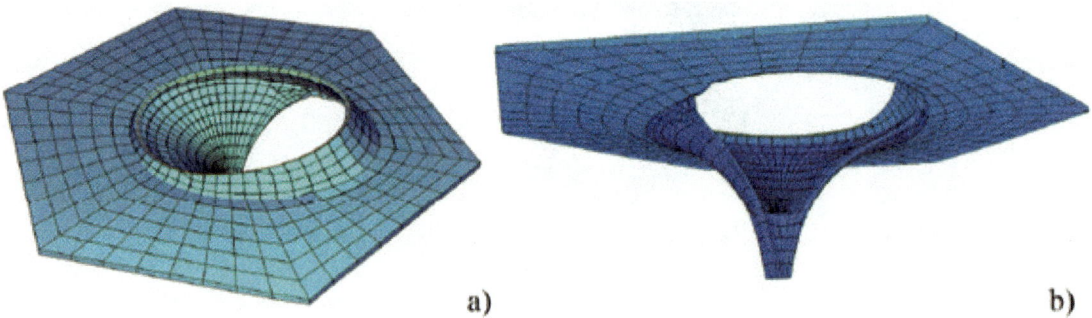

Figure 7.5 Structural model of a chalice. a) View from above, and b) view from below. (Source: Werner Sobek, Stuttgart, 2013.)

to the Mercedes-Benz Museum, which requires a considerable effort in the definition of the rebar geometry. The bars must be bent off-site for the prefabrication, and while most of them are one-directional, a certain portion must be bent in two directions [3].

Nonlinear Analysis and Structural Behavior Optimization

The geometric design protocol of the double-curved roof has been split at level -6m. This allows the upper part of the standard chalices to be described in a way that more or less follows one type of geometry. The different heights at each chalice are mitigated by three types of pedestals that connect the lower end of the chalices with the inclined platforms. Each pedestal has to be adapted to the height of each chalice by sectioning the pedestal at the upper surface of the platform. Through this approach, a high number of geometrically similar chalices were achieved. Several chalices present local changes as they are partly cut at the junction with the sidewalls and at the kinks of the roof. The grouping of the pedestals also leads to a high number of identical surfaces, as they only differ in the lower portion (Figure 7.4).

One of the main challenges in designing and generating various structural models of the railway station is the definition of the middle surface for the concrete shells and the need to obtain a meshed system with continuously varying shell thicknesses. Both of these tasks were solved using Rhino-Scripting, which was also used later in the design stages to define the rebar cage on the basis of a parameter-based optimized design. The fabrication costs were thereby minimized by reducing the number of reinforcement types throughout the structure.

Once the 3-D model progressed toward the first finite element models of the chalice, it showed the entire structural feasibility of the design and the best geometrical optimization results of the geometry. The results were then applied for further form development scenarios until the full finite element model of the station roof was achieved.

The supporting structure of the station hall is a vaulted, seamless concrete double-curved structure. The vault system is divided into 28 modular elements, the chalice supports, between the four platforms and the long outside walls. Each chalice support is hexagonal in plan. The corner points lie in a 40m circle. The upper surface of the roof is even, while underneath the surface is curved. This ensures that the vertical loads in the structure as a whole are transferred mainly via membrane forces and bending.

The geometry of the chalice supports has been developed and continuously optimized using 3-D methods so that all standard chalices are produced with one form. Given the double curvature of the geometry, a counter formwork is necessary. Through the standardization of the 3-D surfaces within the station it is possible to reuse most of the formwork panels. As a result of various feasibility studies relating to the technical requirements, the quality of the surface, and the color, the structure will be made of self-compacting concrete.

Finite Element 3-D Modeling and Automation

The key in setting up a structural model that comprised so many variations of the same typology was the automation of generating the finite element meshes. This was achieved directly in Rhinoceros through scripting (Figure 7.7). The meshed model contained all the information necessary for the finite element analysis, and the data could be exported directly from Rhinoceros into Sofistik by means of text instructions (Figure 7.8). Sofistik was then used as the structural analysis software. The structure was divided into areas that could be better modeled by means of beam elements (lower area of the chalices) and areas that had to be modeled by means of shell elements (upper area of the chalices). The interface between these areas was modeled and checked carefully to assure a correct force transfer.

The differentiation led to the coding of two sets of scripts to derive the model information. One of the key issues for both sets of scripts was the geometrical definition of the center axis, respectively, the middle surface. The center axis was modeled by sectioning the surfaces that describe the lower area of the chalices and by connecting the gravity centers of the sections. The axis geometry and the sections were automatically exported into a text instruction file for further use in Sofistik.

The generation of the middle surface of the shell-meshed regions was more complex. The outer and inner surfaces were preprocessed, defining different subregions to be later analyzed in detail in Sofistik. These surfaces constitute the input for a script that determines the shortest distance between the two surfaces at a discrete number of nodes and within a certain range of approximation. All the necessary information (i.e., element numbering, node thicknesses) was exported into a text file for further use in Sofistik.

This process was developed and checked for one single chalice first and then further implemented to model the whole station, thereby accounting

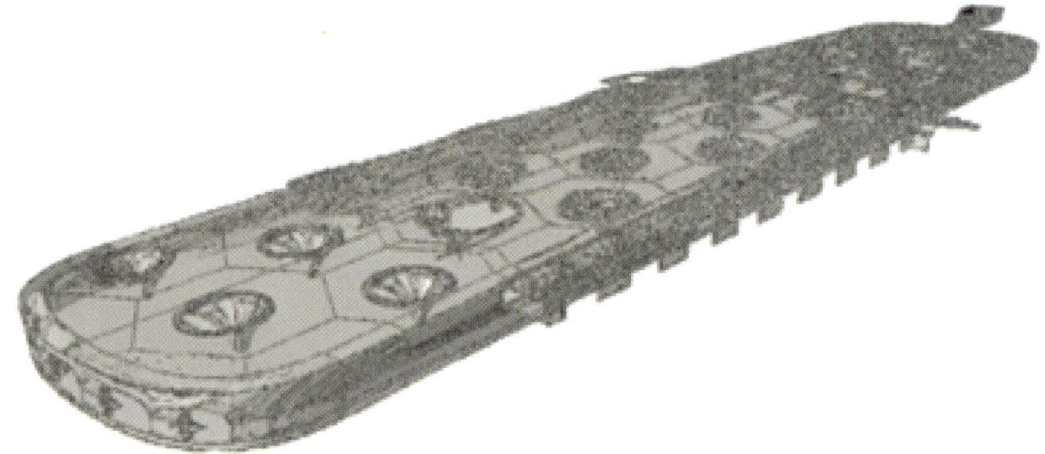

Figure 7.6   Architectural model of the railway station in Rhinoceros. (Source: Werner Sobek, Stuttgart, 2013.)

Figure 7.7   Meshed model of the railway station in Rhinoceros, middle surface. (Source: W. Sobek, Stuttgart, 2013.)

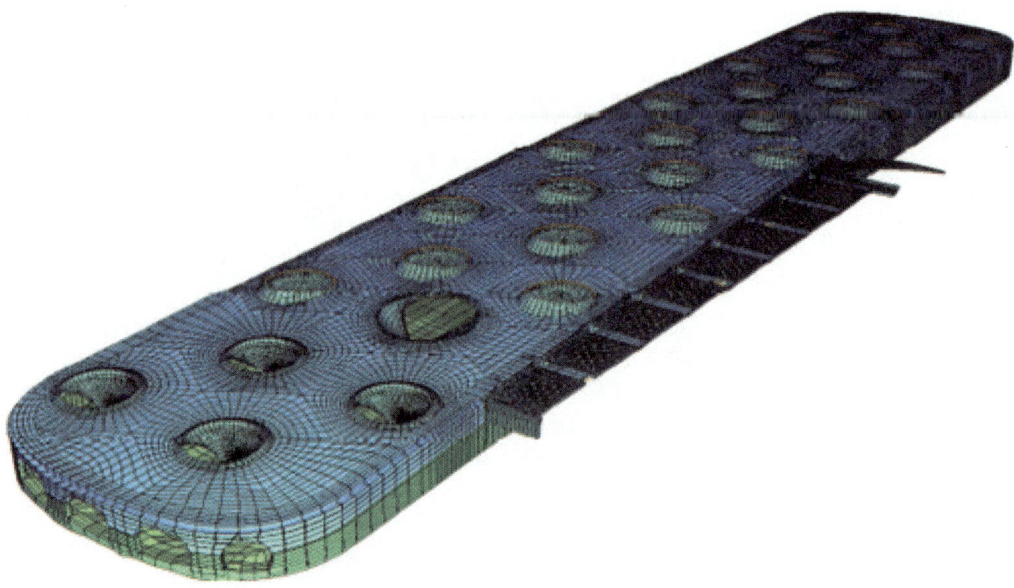

Figure 7.8   Finite element model of the railway station in Sofistik. (Source: Werner Sobek Stuttgart, 2013.)

for all the local variations from the standard geometries. Due to the size of the meshed model, the elements were sorted in groups to better address structural evaluations, rebar topology, construction phases, and other issues. The main rebar directions were also directly defined through scripts in Rhinoceros.

For the Stuttgart 21 project, the checking of the model information became even more important than usual, as the workflow is highly automated. For this reason, the correspondence between the given architectural geometry and the finite element model was checked visually by sectioning every chalice 24 times, as well as analytically by means of data comparison at special sampling nodes. The overall process of structural 3-D-modeling was carried out iteratively multiple times to optimize the railway station structure.

Scripting and Fabrication Process

The architectural 3-D model is the basis for the fabrication of the formwork, with scripting used to define sample points on the surfaces of the formwork for on-site geometric control. The generated sample points and coordinates were inserted in control-point drawings. For the area at the so-called special chalice two sets of control-point drawings had to be generated. The structure in this area has to be precambered, therefore one set of control-point drawings considers the deflections under dead loads, and the second set does not. The definition and optimization of the rebar geometries was a very complex task due to the small radii of curvature of the chalice geometry. The architectural 3-D model was split and tailored into different sets of surfaces to generate the rebar.

This allowed for different reinforcement strategies in consideration of geometric and structural issues. The rebar layout was generated mainly through automated scripting, defined through sets of parameters such as distance to the surface, anchoring length, and rebar distance.

As the bars are generated based on doubly curved surfaces, the resulting

Figure 7.9   3-D model of the rebar's cage produced parametrically. (Source: Werner Sobek, Stuttgart, 2013.)

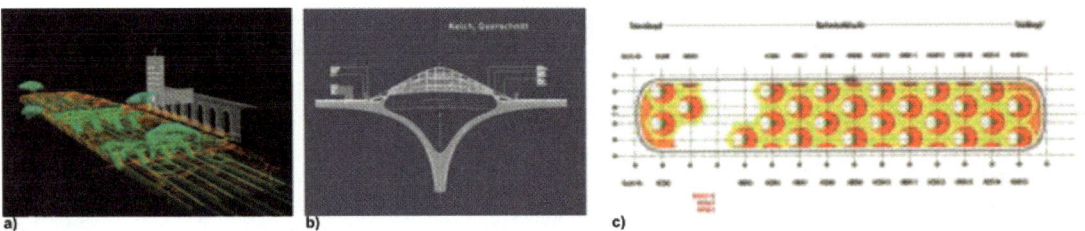

Figure 7.10 a) Surface model of the railway station in Mircostation. b) 3-D generated cross section of a chalice. c) Construction schedule of the chalice elements. (Image courtesy of ingenhoven architects.)

geometries are NURBS splines. Scripts were developed to convert these splines into polylines made of straight segments or arch segments with similar rebar geometries grouped together within certain tolerances. Fabrication issues such as segment lengths and the deviation between the original and converted geometries were also verified using computer scripts, thus allowing for further optimization and control.

The overall design process has been set up with an automated workflow, making use of specially developed algorithms to allow an iterative structural optimization.

During the design development phase, the main challenge was the generation of a detailed finite element model with constantly varying thickness based on the architectural 3-D model. Later on, during the preparation of the construction documents, the definition of curved or polygonal rebars, matching the steady change of given surface curvatures, was particularly challenging (Figure 7.9). In this case, it was fundamental to consider the rebar production limits and strive to keep the costs within a reasonable range. Overall, the use of scripting has proven to be a very powerful tool by allowing automated changes in the modeling work and in optimizing the workflow between the different professionals involved [3].

Assembly Process

The entire prefabricated construction of the station hall will be erected in several sections, both for water supply and distribution reasons. Moreover creep constraints in the concrete are avoided by defining a proper construction sequence.

Low Primary Energy Requirements

The engineers and architects tried to keep the cumulative primary energy consumption profile for the selected building materials of the station hall during manufacture as low as possible. Concrete has a primary energy factor of approx. 400 kWh/m^3. The steel for the reinforcement bars and the concrete has been reduced to a structural minimum, with the result that the concrete shell of the station roof has a primary energy factor of around 200 kWh/m^2. By comparison, a theoretical alternative structure for a station hall made of steel, concrete, and glass would have a primary energy factor of approximately 5,000 kWh/m^3, (i.e., energy consumption would be around 25 times higher) [4].

Zero-Energy Station and Passenger Comfort

The Stuttgart station will be the prototype of a new generation of railway stations that will provide passenger comfort with passive strategies on the highest level. The parametric-algorithmic simulation results with Bentley Hevacomp Microstation and with computational fluid dynamics (CFD) substantiates the thermal comfort requirements of the obligatory air velocity and temperature profiles for a zero-fossil-energy station, which requires in this case no mechanical heating, cooling, or ventilation and it must provide enough lighting in daytime while avoiding CO_2 emissions.

The basic ecological concept makes use of passive and active hybrid strategies using natural energy resources to operate the building and to provide thermal comfort through the year. This is accomplished through the geothermal use of the physical cold and heat storage mechanisms of the earth. The natural ventilation of the station is provided through the train and passenger circulation tunnels in conjunction with the operable and automated natural light eyes of the chalices. The simulated natural ventilation rate provided by train-induced flow in connection with thermal flows creates a maximum air velocity of 1.0–1.5 m/sec on the platforms.

The high level of thermal comfort, which is achieved without any input of thermal energy, is reflected in the average perceived temperatures on the platforms of 20°C–22°C in summer and 5°C–8°C in winter. By cutting back to what is necessary, a high level of comfort was achieved for extremely low primary energy consumption. The expected temperatures in the tunnels will vary in the course of the year by an average of +10°C and, given the flow conditions, rarely reach temperatures above +20°C in summer or 0°C in winter. All the passive and active hybrid systems are dynamically controlled through an intelligent building automation system.

The integrated light eyes are evenly distributed through the vaults and openings above the platforms, guaranteeing that the station hall has an adequate supply of natural light for up to 14 hours a day. The conducted lighting simulations confirm that the modeled and selected light surfaces ensure that the interior of the station is pleasantly lit even during overcast weather. At night, the underside of the shell structure is used to reflect the artificial light-emitting diode (LED) lighting technologies.

The station, which is lit with natural light and has an average daylight quota of over 4%, is equipped with an auxiliary renewable energy system to back up the natural light. A maximum specific connected load of 3 – 9 W/m^2 is achieved by varying the ratio of direct and indirect light input [4].

Participants
Client: Deutsche Bahn AG, Berlin; represented by DB Projekt Bau GmbH Niederlassung Südwest, Projektzentrum Stuttgart 21ThyssenKrupp AG, Essen, Germany
Architects: Ingenhoven architects, Düsseldorf, Germany: Christoph Ingenhoven, Klaus Frankenheim, Hinrich Schumacher, Dieter Henze
General Planning: Stefan Höher, Annemone Ingenhoven-Feld, Christian Kawe, Peter Pistorius, Bjørn Polzin, Frank Reineke, Marc Böhnke, Barbara Bruder, Ralf Dorsch-Rüter, Peter Jan van Ouwerkerk, Alexander Prang, Michael Reiß, Takeshi Semba, Maximo Victoria, Tom Wendlinger, Harald Wennemar, Lutz Büsing
Location: Stuttgart, Germany

Area: 185,000 m²
Building Costs: 250 Mio. Euro
Start and Completion: 2013–TBD
Scientific Support for Form-Finding, Design Consultant: Prof. Dr. Frei Otto, Leonberg
Construction and Structure
Werner Sobek Stuttgart, Germany (from Schematic Design to Construction Documents) and Tragwerksplanung S21 Hauptbahnhof GbR, Leonhardt Andrä und Partner, Stuttgart mit Happold Ingenieurbüro, Berlin (Concept Design)
Façade Consultant: DS-Plan, Stuttgart, Werner Sobek, Stuttgart
Building Services: DS-Plan, Stuttgart HL-Technik AG Beratende Ingenieure, Frankfurt a.M., Kofler Energies, Frankfurt a.M
Building Physics: DS-Plan, Stuttgart
Fire Protection: BPK Brandschutz Planung Klingsch, Düsseldorf
Ventilation Analysis: IFI Institut für Industrieaerodynamik, Aachen
Artificial and Natural Lighting: Tropp Lighting Design, Weilheim
Landscape Architecture: Ingenhoven architects, Düsseldorf, WKM Weber Klein Maas Landschaftsarchitekten, Meerbusch
Traffic Engineering: Ingenieursgruppe für Verkehrsplanung und Verfahrenstechnik, Aachen, Durth Roos Consulting, Darmstadt
Project Management: Drees & Sommer Infra Consult & Management, Stuttgart

References

[1] Interview with Michael Reiß, Peter Pistorius, Martin Reuter of ingenhoven architects in Düsseldorf on June 15, 2013.

[2] Werner Sobek, Dietmar Klein, Thomas Winterstetter. Hochkomplexe Geometrie. Das neue Mercedes-Benz-Museum in Stuttgart (VDI-Beratende Ingenieure, 2005). 10:16–21.

[3] Lucio Blandini, Albert Schuster, Werner Sobek. The railway station "Stuttgart 21" structural modeling and fabrication of double curved concrete surfaces, in Proceedings of the Design Modelling Symposium Berlin 2011, pp. 217–224. (C. Gengagel, A. Kilian, N. Palz, and F. Scheurer. Springer Edition).

[4] Global Holcim Awards 2006, http://download.holcimfoundation.org/1/docs/DE_booklet.pdf, accessed on August 8, 2013.

Chapter 8

Integrated Project Delivery and Total Building Automation: Q1 ThyssenKrupp Headquarter, JSWD Architekten + Chaix & Morel et Associés

Thomas Spiegelhalter

Introduction

ThyssenKrupp is a global technology group with employees in more than 80 countries. In 2006, the ThyssenKrupp AG with ECE project management in Germany hosted an international two-phase architectural competition for the rehabilitation of the former Friedrich Krupp brownfield area. Approximately 100 offices submitted their competition designs, and finally in the second stage the team of Chaix & Morel et Associés from Paris, France, and JSWD Architekten from Cologne, Germany, won the first prize for the commission to design and build the approximately 17-hectares Krupp-ribbon campus. The second prize was awarded to Brüning Klapp Rein, Essen, Germany, and the third prize to Zaha Hadid Architects Ltd., London. The expectations of the clients were focused on dynamic space structures that would flexibly react to changes within the company of more than 2,000 employees (Figure 8.1).

The following case study will focus on the successful integrated project delivery process of the Q1 ThyssenKrupp Quarter in Essen, Germany. This cube-shaped, corporate, nearly zero-fossil-energy headquarter Q1 forms the centerpiece of the campus with ten other striking buildings from Q2, the forum, and two seven-story administrative buildings Q5 and Q7. All 170,000 m² large buildings including underground garages and parking structures, were designed, built from 2009 to 2010, and decommissioned from 2011 to 2012 [1].

The standard approach, originating in sketches, analog models, 2-D plans and sections and then proceeding into technical detail design—a linear approach—receded as the architects began to conceive the Q1 project in nonlinear 3-D digital model workflows. This methodology allowed the architects, engineers, companies, and the client to share, visually present

Figure 8.1 Site plan ThyssenKrupp Quarter, JSWD Architekten and Chaix & Morel et Associés, 2008, photo model competition ThyssenKrupp Quarter, ThyssenKrupp AG, 2006. (Image courtesy of JWSD Architekten, 2013.)

Figure 8.2 Immersive Open GL Performer Simulation. (Images courtesy of Guenther Wenzel, Virtual Environments, Institute for Human Factors and Technology Management (IAT) at the University of Stuttgart, 2013.)

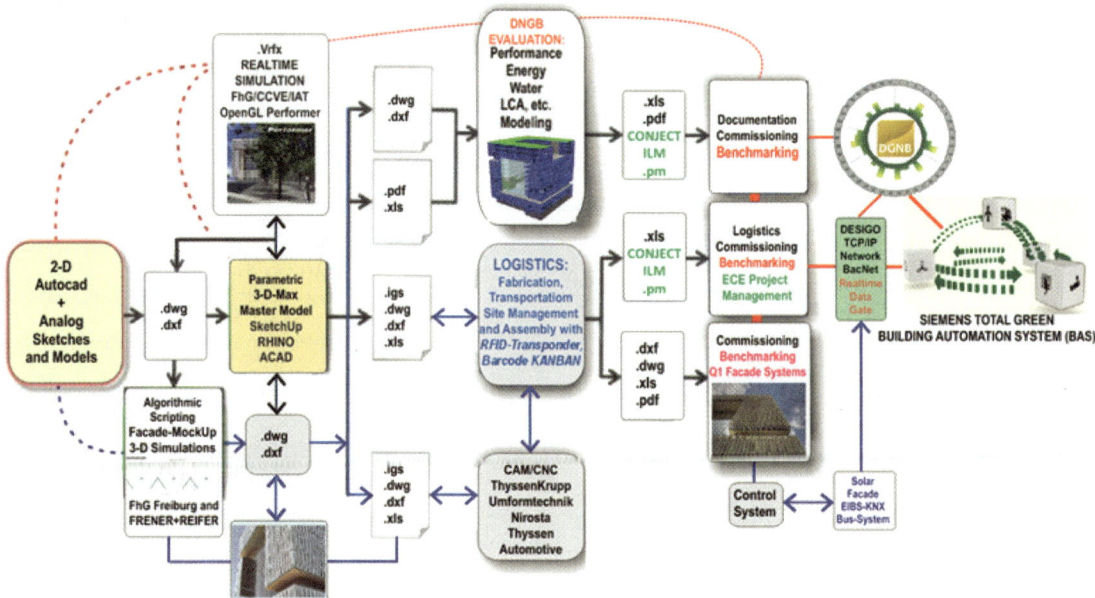

Figure 8.3 Schematic workflow diagram of the integrated project delivery for Q1 ThyssenKrupp Quarter from the early design stage, 3-D-simulations, detailing, file-to-production process, assembly, commissioning, benchmarking, and total green building automation. (Diagram courtesy of Thomas Spiegelhalter, 2013.)

in real-time and discuss all the parametric data input in varying degrees of specificity of parts, zonings, and regions.

According to the lead architect Jürgen Steffens of JSWD Architekten, "the computational shift into 3-D also paralleled and engaged a conceptual and pragmatic shift in the office, a circumstance which was required to navigate and manage through increased organizational complexity and required flexibility of the 300,000 m² gross floor area with 10 different occupancy types and zonings" [2].

According to Steffens, the first shift in the design process from 2-D AutoCAD / SketchUp to 3-D-Max and Rhino/Grasshopper happened when the client demanded a more collaborative and transparent working environment. The second shift occurred when the complexity of the project demanded real-time 3-D simulations with VRfx-compatible formats in OpenGL Performer software to share, reiterate, synchronize, and visualize quick changes and updates in the parameters with the client and special contractor direct input [2].

The VRfx open format of the OpenGL Performer was developed by the German. The Fraunhofer Gesellschaft (FhG) in collaboration with the Competence Center Virtual Environments (CCVE) of the Fraunhofer IAO and the Institute for Human Factors and Technology Management (IAT) of the University of Stuttgart [3].

The spatial 3-D visualization for direct interaction and immersion into the virtual architecture of Q1 was facilitated in particular by Caetano da Silva and Prof. Guenther Wenzel from Fraunhofer IAO in Stuttgart for JWSD Architekten and Chaix & Morel et Associés [4]. It was clear at this level of collaborative design that the virtual reality tool could be simultaneously a medium, prototype, and rapid modeling tool to visualize, manipulate data and analyze complex relationships intuitively.

The Fraunhofer Visual Processing of Three-Dimensional Content (VRfx) format has similarities with other 3-D game engines, such as Quest3D® fed by 3-D-data resulting from multiple scenario inputs of built environment modeling (BEM) tools, to create navigable, immersive, integrated 3-D architecture models or urban environments using false color.

According to Guenter Wenzel "These virtual models enhance designers and clients" perception by allowing them to explore experientially hundreds of potential multidisciplinary design combinations, both qualitatively and quantitatively, in a practical, collaborative and cost effective way [4]. Participatory processes with accessible data exchange are essential for getting necessary change and radical design innovations adopted. Today, designers have the opportunity to create interactive models with interrelated data that are immersive—including a mix of real and symbolic representation—that facilitate experiential design and 'what if'-scenarios' (Figures 8.2 and 8.3).

Q1 Sun and Daylight Prototype Control System: Design and Production

Sun and daylight control systems with suitable glare remediation while still giving the building occupants sufficient out-views and visibility require an integrated design approach. The architects designed successfully a direct link to the dynamic and perpetually evolving patterns of outdoor illumination and direct solar radiation with a unique controlled but dynamic and adaptive sensor controlled daylighting façade system. It is beneficial clearly helps to create a visually stimulating and productive environment for the building occupants while also reducing significantly the total building energy costs and greenhouse gas emissions (Figure 8.4).

All the Q1 office façades have slim steel profiles and operable vertical lamellas. In total there are 400,000 stainless steel slats covering an area of approximately 8,000 m² to control sun and daylight entering the building. This prototype system remains fully functional even in high wind speeds of 70 km/hour, which is unusual for high-rise buildings. Most outside

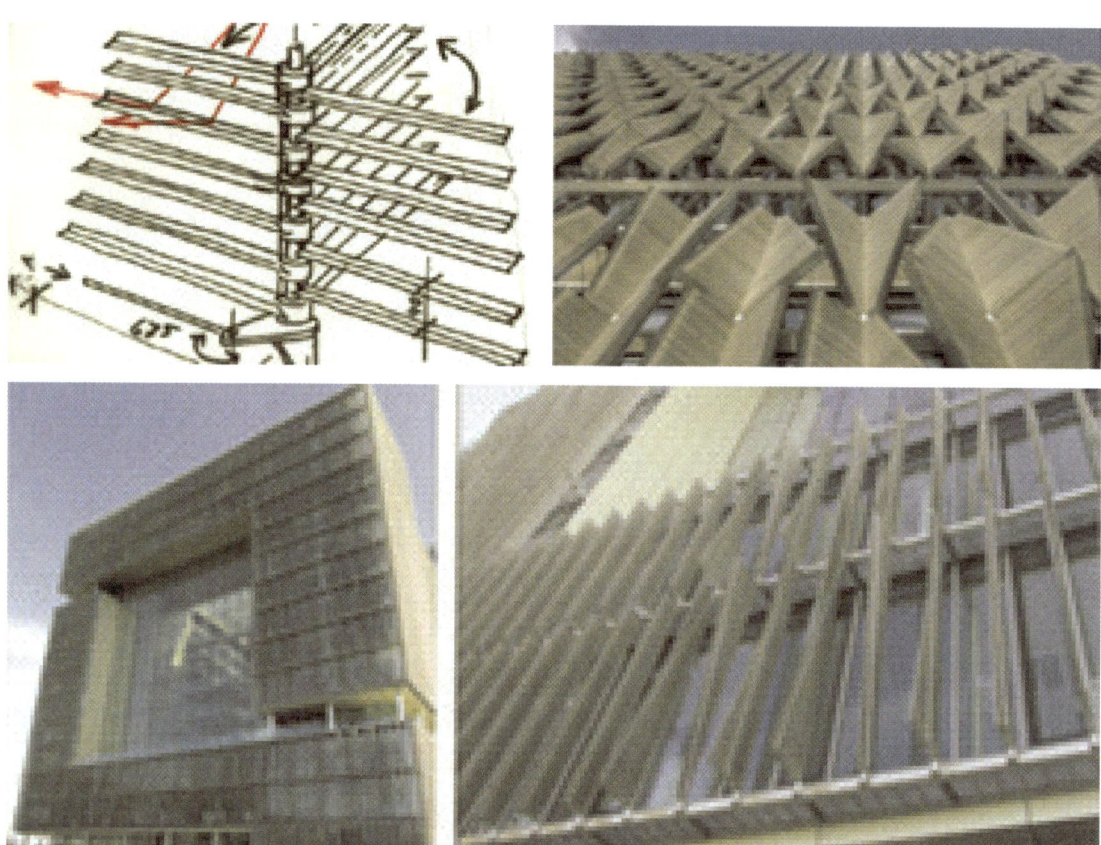

Figure 8.4 Façade systems of the Q1 ThyssenKrupp quarter, Essen, Germany. Planning and design: JSWD Architekten and Chaix and Morel et Associé with Frener and Reifer and the Fraunhofer Institute for Solar Energy Systems. (Images courtesy of Thomas Spiegelhalter and Günter Wett / Frener and Reifer, 2010.)

shading and daylight systems in similar building types around the world turn off during the high-speed wind season, and as a result their energy consumption increases dramatically due higher cooling loads. The solar and daylighting control concept was schematically designed by JSWD Architekten and Chaix & Morel et Associés and algorithmically detailed and optimized through multiple simulations and digital mock-ups (DMUs) by the Fraunhofer Institute for Solar Energy Systems (FhG) in Freiburg, Germany in collaboration with the special façade contractor Frener+Reifer from South Tyrol in South Tirol, Italy. The DMUs included a parametric-algorithmic dynamic weather data-driven solar radiation and illumination analysis with multizone simulations to engineer a constant horizontal overhang for summer sun protection and daylight control. All systems were combined with a vertical set of twisting fins. The matte underside of the system was further designed by the architects and engineers and Frener+Reifer after the FhG modeled and optimized various property behaviors such as reflection, shadow, daylight, glare, and heat protection (Figure 8.5) [6].

All the shading elements are sensor controlled through a separate EIBS-KNX-Bus system and embedded for data recording in the SIEMENS BAS with the assistance of 1,600 electric motors to dynamically perform to the following settings:

- Closed or parallel to the glass façade

- Moving with the sun / variable and perpendicular to the sun
- Open: the horizontal lamellas move perpendicular to the glass façade (Figure 8.5) [7]

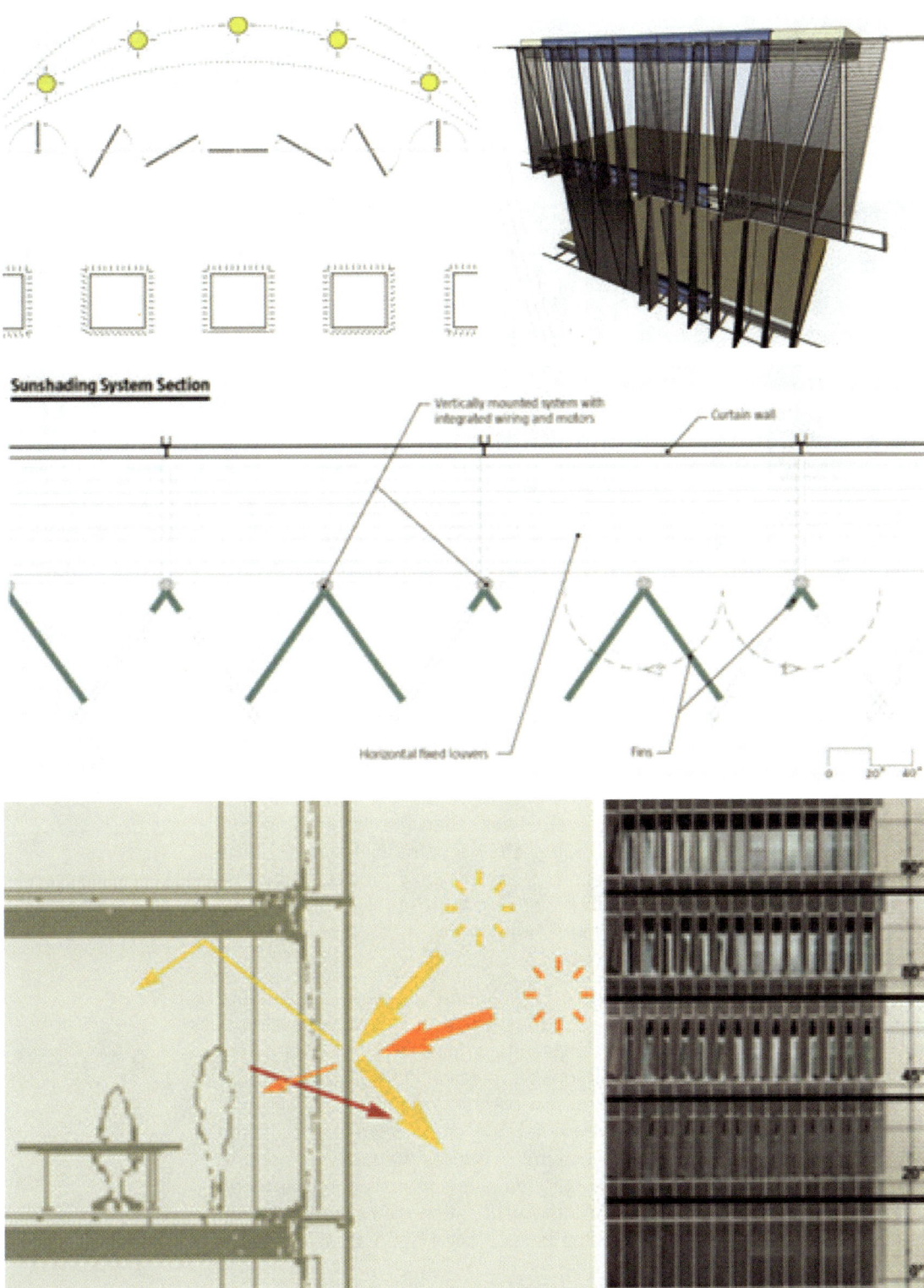

Figure 8.5 Q1 lamella performance angles perpendicular to the glass façade. Image right, bottom: The virtual animation of the façade by Frener + Reifer. (Images courtesy of JSWD Architekten and Chaix & Morel et Associés, 2013.)

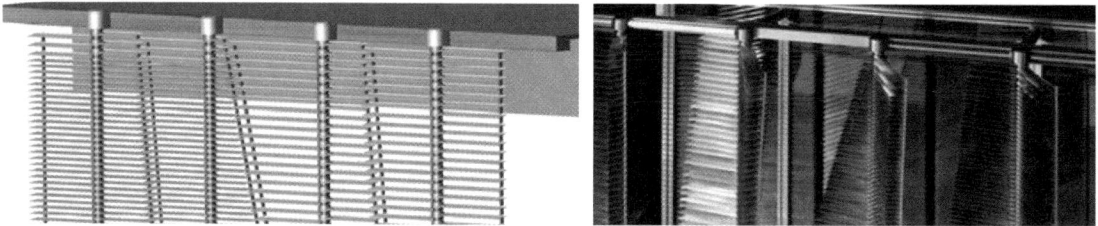

Figure 8.6 Left: Sun protection and daylight control system concept with digital 3-D mock-ups (file formats: ACAD, .dwg, dxf., .3ds). Right: Analog façade design variation and functional prototype mock up. (Source: Frener+Reifer, Sept. 2013.)

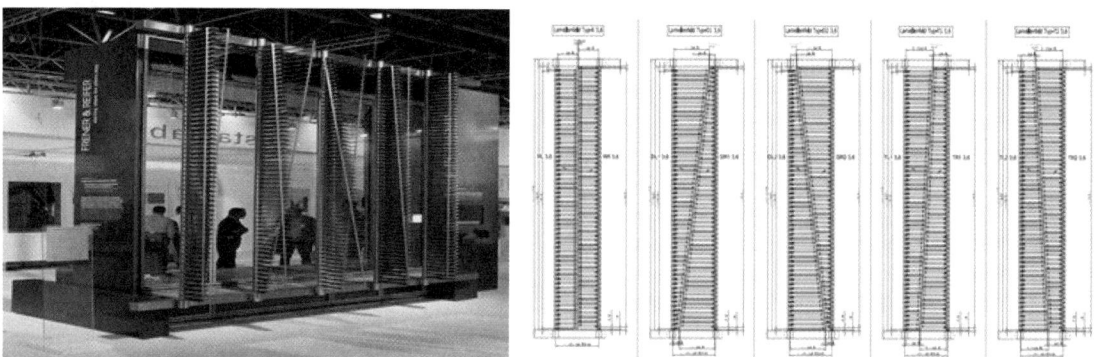

Figure 8.7 Left: Q1 fully functional series solar control and thermo façade prototype mock up with integrated control system. Right: Part and component variant system, based on logistic concept. (Source: Frener+Reifer, 2013.)

The geometrical differentiation into trapezoid, triangular, and rectangular elements creates a façade that follows the sun and controls the daylight for efficiently illuminating the interior.

The preliminary design-to-factory-file process of the innovative façade system was simulated and a 3-D mock-up tested by the Fraunhofer Institute of Solar Energy (ISE) and the special façade contractor Frener and Reifer from South Tyrol in Italy under the supervision of General Planner ECE and the client ThyssenKrupp Real Estate. Then the different analog and digital file formats for the Q1 value chain control batch of 22,000 lamellas were tested, cleaned, calibrated, and digitally processed by the Simultaneous System Engineering group of ThyssenKrupp Umformtechnik and ThyssenKrupp Nirosta, the group's automotive manufacturing units (Figures 8.6, 8.7, and 8.8) [7].

In general, the ThyssenKrupp System Engineering workflow contains an integrated project process and system management for automotive systems and parts. This includes all project phases beginning with the parametric concept study through to production design, physical and digital mock-up (DMU) testing, assembly, and production planning—in this case—to the façade system and component delivery for Q1. The integration of testing within the early phase of the development is predominantly important for the correlation of multidimensional simulation models through finite element methods and contributes to the verification of the computation results before any fabrication planning takes place (Figures 8.6, 8.7, and 8.8).

Frener+Reifer, Thyssen Krupp Umformtechnik, and Nirosta used for the logistics from fabrication and transportation to assembly a barcode-

Figure 8.8 Left: Structural analysis with power train variation prototype. Middle: Part and component variant system, based on logistic concept. Right: Wind tunnel and flash testing. (Image courtesy of Frener+Reifer, 2013.)

Figure 8.9 Barcode controlled JIT supply chain production using the integrated KANBAN supermarket system with Excel data sheets for logistics from fabrication and transportation to assembly. Optimization protocols with radio frequency identification (RFID). (Images courtesy of Michael Reifer, Frener+Reifer, 2013.)

controlled just-in-time (JIT) supply chain production such as the integrated KANBAN supermarket system and optimization protocols with RFID-radio frequency identification (RFID). RFIDs are mainly used to monitor, identify, and control automated production, transportation, assembly, and positioning, and they also help to avoid product piracy because all parts have encrypted codes and chips (Figure 8.9) [7].

After the modular façade systems with each composed four lamella leaf trees including the power train with the automation components have been manufactured, preassembled, and tested, the façade contractor Frener+Reifer then assembled and mounted on-site approximately 160 slats onto each axis to create electrically sensor-driven slat packages. During this process, it was important that the slats remained movable in the center axis and reacted precisely to the signals of the electrical drive and sensor control system of the EIBS-KNX-Bus that was connected to the central data gate to the SIEMENS building bus control and monitoring the SIEMENS BAS [7]. The programming of the sensor-driven control system includes real-time daylight provision in the interior, glare control, seasonal sun position, and the weather patterns assisted through the weather station on the roof. The fully automated façade system is separated with a parallel

Figure 8.10 Q1 lamella performance angles perpendicular to the glass façade. (Images courtesy of ThyssenKrupp AG, Germany, 2013.)

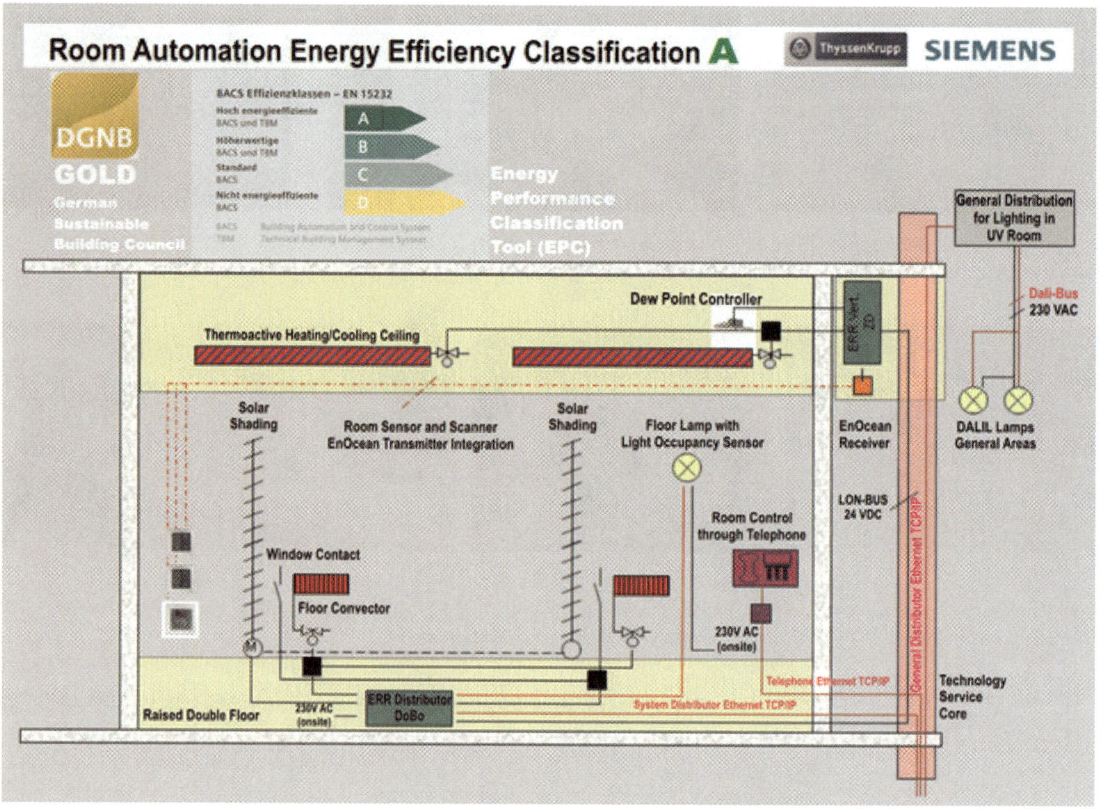

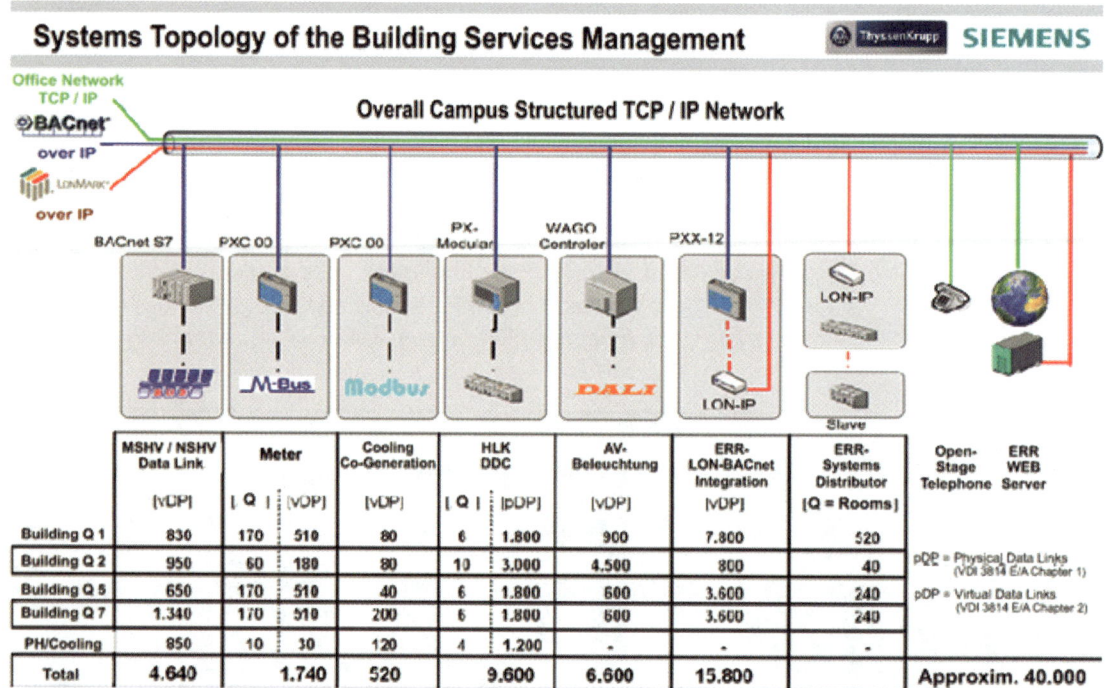

Figure 8.11 Room automation for solar shading, daylight control, artificial lighting, heating, and cooling, dew point control, benchmarked with Energy Efficiency Classification A according to the BACS European Norm (EN) 15232. Diagram adapted and translated by Thomas Spiegelhalter. (Source: Hubert Dierkes, Siemens Infrastructure & Cities Sector Building Technologies Division, Essen, Germany, 2013.)

Figure 8.12 Overall systems topology of the building services management for all campus buildings Q1, 2, 5, and 7. Diagram adapted and translated by Thomas Spiegelhalter. (Source: Hubert Dierkes, Siemens Infrastructure & Cities Sector Building Technologies Division, Essen, Germany, 2013.)

bus system from the overall Siemens BAS and energy control system, coupled in a central control gate to minimize the risk of collisions of the sun protection lamella components and potential data package overloads that could occur with direct decentralize data package integration into the overall BAS control system (Figures 8.3, 8.11, and 8.12).

Based on the previous algorithmically developed scenario simulations and detail recommendations for the sensor baseline protocols by the Fraunhofer Institute, all the slats of the façade system turn outwards so that the sun shades remain open on cloudy days. Even when the slats are closed directly in front of the façade, employees can open the windows and access for maintenance is always possible. The façade contractor Frener and Reifer installed the inner curtain wall with German Schüco System elements. Both skins in Q1, curtain wall, and fins are approximately 8,000 m² each.

Q1 Energy Concept

The ThyssenKrupp Quarter's primary energy consumption of 139.6 kWh/m²/yr is 58% lower than the legally allowed value of 242.3 kWh/m²/yr of a similar reference. Its ecological footprint is characterized by 27% less CO2 emissions than similar buildings in this category. The remaining energy need, for the 139.6 kWh/m²/yr, is matched through a highly effective geothermal system with an underground heat and cold storage and a waste heat energy recovery system. Renewable geothermal heating and cooling is supported by an installation of a 1,000 m² geothermal field of ground loops at depths of up to 100 m. The geothermal system not only heats and cools the buildings, it can also be used to store surplus heat or cold in the ground during every season. In combination with this technology, all concrete floor slabs are operated with integrated thermally activated cooling and heating flex pipe systems. The thermal comfort room temperature is set between 21°C and 26°C. Heat is continuously recovered from waste air via the office's central ventilation heat recovery system, which is then converted and reused through absorption and cogeneration technology [6].

General Q1 Project Management

The overall planning with a compatible 3-D master model for all the project participating architects, engineers, and more than 300 companies and 50 involved planning and expert offices was coordinated by ECE Projekt Management GmbH & Co. KG, in Hamburg, Germany. ECE as general planner and site supervisor was mainly responsible for the organization of the international competition in 2006 and the organization of overall design and the construction process including commissioning of the ThyssenKrupp Quarter until its completion in 2010.

The management of the total material inventory contained:

- 23,000 metric tons of steel
- 90,000 m³ of concrete
- Approximately 16,000 m² of glass
- 400,000 stainless steel sunshade slats
- 320,000 m of electric cable

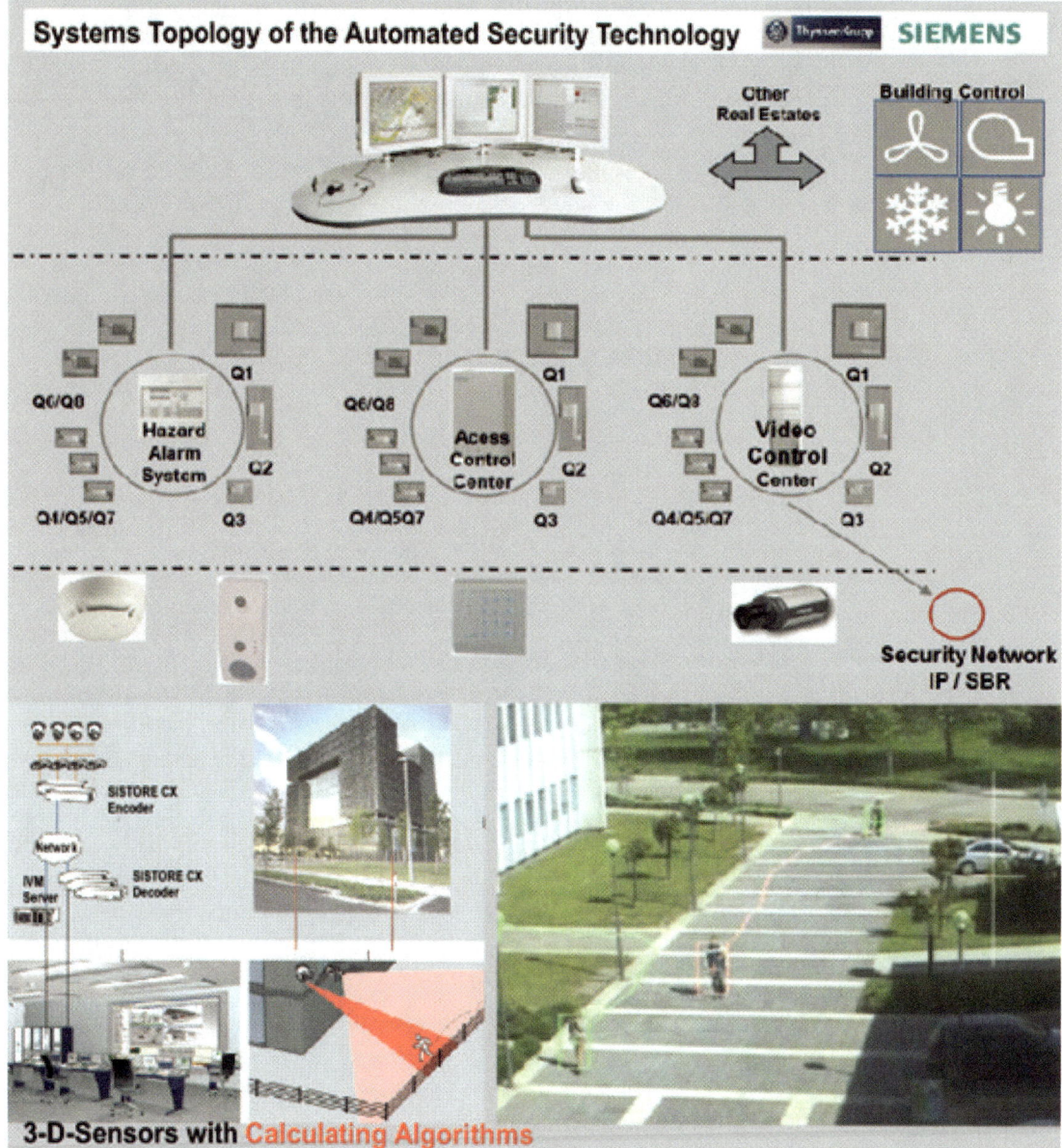

Figure 8.13 Fire, security, and building control management system with calculating algorithms. The security management system coordinates 350 video surveillance cameras, 490 card readers, and a 3-D intrusion detection system equipped with 2,750 intrusion detectors. Diagram adapted and translated by Thomas Spiegelhalter. (Source: Hubert Dierkes, Siemens Infrastructure & Cities Sector Building Technologies Division, Essen, Germany, 2013.)

- Approximately 9,000 m of water piping
- Approximately 10,000 m² of parquet flooring
- Approximately 30,000 m² of carpet

By organizing the immense amount of information in a parametric 2-D/3-D model, thousands of detailed plans had to be generated automatically for any variant of the design chosen by the architects. The parametric models captured those persistent rules behind the developing form, reducing thousands of coordinates to a handful of standardized parameters individually expressed in variables and protocols.

ECE used the Internet-based CONJECT information life-cycle management

(ILM) data platform to cover all processes throughout the life-cycle of the real state, from the planning and construction through to the management and usage of the real estate, encompassing both the technical as well as financial aspects [10].

Q1 Total Green Building Automation (DESIGO)

Siemens specifically developed a BAS called Green Building Monitor™ to supervise the building consumption data of Q1 continuously for expert analysis [11]. The BAS includes a specially developed energy-efficient operating control concept for 1,133 rooms. Sensors record daylight and detect the presence of people, which allows the system to provide optimal workspace conditions based on those readings. This scheme does not include traditional switches because telephones are used to customize all functions for each workspace. The system displays accurately to the hour how much power and water is consumed in the building, the temperature, and greenhouse gas emissions. The BAS monitors user behavior and compares, measures, and visualizes real-time progress data with various performance benchmarks for carbon neutrality or the nearly-zero fossil energy standards. Other features include integrated control for electrical and security technology tailored to the specific needs of the ThyssenKrupp campus buildings [12].

The security management system coordinates 350 video surveillance cameras, 490 card readers, and an intrusion detection system equipped with 2,750 intrusion detectors. In addition, Siemens introduced a new solution that provides for touchless, video-based identification and counting of the Q1 populace. The security concept consists of a fire alarm system including aspirating smoke detectors and voice alarm.

Q1 Certifications, Awards, and Honors

During the design and planning process and upon completion, the building was evaluated by the auditor and engineer Gerhard Hoffmann, Ifes GmbH, Frechen, in Germany. It then was certified with the highest German Sustainable Building Council's (DNGB) Gold Certificate for its successful Integrated Practice of ecological and economical building systems, materials, and sustainable operation management. For the calculation of the life-cycle assessment (LCA) and eco-balancing, the auditor used the international environmental product declaration (EPD) standard procedure

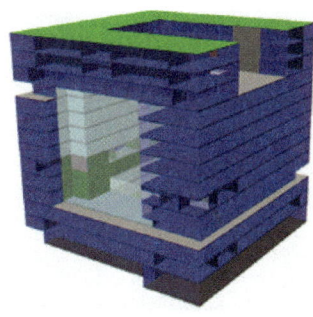 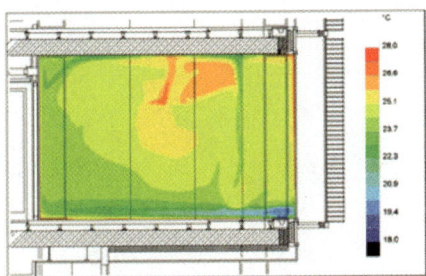 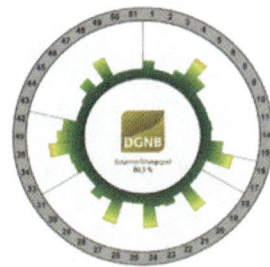

Figure 8.14 Left: Evaluation excerpt: thermal zoning model. Middle: Summer thermal comfort office model. Right: The German Sustainable Building Council's (DNGB) Gold Certificate evaluation diagram for the Q1 ThyssenKrupp building. (Images courtesy of Gerhard Hoffmann, Ifes GmbH, Germany, 2013.)

(Figure 8.14). The building has also received the LEAF Award 2011 for the Best Sustainable Technology Incorporated Into a Building 2011, the BDA Essen Award, the 2012 German Steel Construction Award, and the Iconic Awards 2013 for best Corporate Architecture, in addition to others.

Project Participants

Client: ThyssenKrupp AG, Essen, Germany
Architects: JSWD Architekten, Cologne, Germany and Chaix & Morel et Associés, Paris, France
Location: ThyssenKrupp Allee 1, 45143 Essen, Germany
Area: 170,000 sqm
Year Completed: 2010
Client: ThyssenKrupp AG
Exterior Facilities: Kipar Landschaftsarchitekten, Duisburg, Germany
Façade Consultants: Priedemann Fassadenberatung, Berlin and Werner Sobek Ingenieure GmbH & Co. KG, Stuttgart, Germany
Structural Engineering: IDN Ingenieurbüro Domke Nachfolger GbR, Duisburg, Germany
Lighting Design: Licht Kunst Licht AG, Bonn, Germany
Construction Sunshade and Daylight Façade: Frener and Reifer GmbH, Brixen / Bressanone, Italy

References

[1] ThyssenKrupp 2011. "ThyssenKrupp Quarter: JSWD Architekten Chaix & Morel et Associés." Jovis; Bilingual edition (April 30, 2011).

[2] Interview with Juergen Steffens, JSWD Architekten, Cologne, Germany, on March 13, 2013.

[3] Institute for Human Factors and Technology Management (IAT) of the University of Stuttgart, Germany. http://www.iao.fraunhofer.de/lang-en/, accessed on March 23, 2013.

[4] Interview with Prof. Dr. Guenther Wenzel from FRAUNHOFER IAO (Institut fuer Arbeitswirtschaft und Organisation), on March 18, 2013.

[5] Thomas Spiegelhalter. Mitigate, Adapt, Sustain: Emerging Workflows and Design Protocols for Subtropical Carbon-Neutral H2 Cities, 2013 ACSA FALL CONFERENCE DESIGN INTERVENTIONS FOR CHANGING CLIMATES (Oct. 2013).

[6] ThyssenKrupp 2011. "ThyssenKrupp Magazine—Architecture. www.thyssenkrupp.com. Accessed on Dec. 23, 2012.

[7] Frener&Reifer: http://www.frener-reifer.com/projekte/thyssenkrupp-quartier-hauptverwaltung/. Accessed on Sept. 15, 2013.

[8] Interview with Michael Reifer, Frener+Reifer, Brixen-Bressanone (BZ), Italy, on September 27, 2013.

[9] ThyssenKrupp Umformtechnik, http://www.tka-as.thyssenkrupp.com/en/services/engineering.html. Accessed on Dec. 23, 2012.

[10] CONJECT, http://www.conject.com/at/en/index. Accessed on Sept.

15, 2013.

[11] SIEMENS DESIGO Building Automation, http://www.buildingtechnologies.siemens.com/bt/global/en/buildingautomation-hvac/building-automation/building-automation-and-control-system-europe-desigo/Pages/desigo.aspx. Accessed on Sept. 15, 2013.

[12] Interview with Hubert Dierkes, Siemens AG, Infrastructure & Cities Sector Building Technologies Division, Germany, on June 15, 2013.

Chapter 9

Design Computation at Arup

Clayton Binkley, Paul Jeffries, and Mathew Vola

Introduction

Design computation is a natural process for engineers—the fundamental nature of our practice is to take abstract design goals and solidify them within a mathematical and physical framework. We apply codes, algorithms, and equations and blend them with intuition in order to realize physical objects.

Engineering practice in the construction industry has shifted over the last 20 years from being a manual occupation with digital support to one in which content is fundamentally intertwined with computation. The process of analysis and documentation is now wholeheartedly engaged. This shift is a function of many factors: cheap computing power and data storage, sophisticated analytical tools appearing in everyday life, and the low barrier to software development. As a result, design computation—that is, custom software and algorithm development in the service of design—has become common practice at Arup. A colleague likes to use the phrase algorithmic people to describe the new generation of engineers who marry design intuition and the physical act of construction with the systems-based approach of computational practice.

Mass customization in design has obviously not been limited to the designer's desktop; similar advancements have been happening on the fabrication side of the industry as the cost of CNC machinery and CAD/CAM software has fallen dramatically and its ease of use and control sophistication continue to rise. More exciting than the open-ended formal possibilities are the opportunities available for designers to access the digital fabrication process, interacting directly with the fabrication models

Figure 9.1   National Maritime Museum of China. (Image courtesy of Cox Architects.)

and CAM software APIs to convert design data directly to fabrication code. New team structures and contractual arrangements allow direct design data exchange between the design and construction teams at all phases of the project.

Moving beyond the static, physical domain of construction, now buildings, infrastructure, and even construction sites are wired and available for interaction. Arup has developed software to interact with the built environment, from excavation monitoring systems at the Transbay Transit Center excavation to Newport Beach City Hall's desktop notifications that guide occupant behavior and achieve optimal natural ventilation system performance.

We are constantly reenvisioning how we approach design and over the past years have recognized the fundamental role that software development and computation play in our practice, spanning site investigation, advanced analysis, data processing, visualization, documentation, and immersive environments, monitoring, and construction. We have chosen to highlight three working areas relevant to current thinking about design computation: parametric design, custom tool development, and automation/optimization.

Parametric Design Case Study

The National Maritime Museum of China consists of five hall structures radiating out to the harbor and converging in a central preface hall. The fluid forms of the pavilions are designed to evoke a wide variety of maritime interpretations without obvious references: rolling waves, an outstretched hand over water, coral or a sea anemone, a small fleet of moored vessels in port, or perhaps a school of jumping carp. Large cantilevers at the building extremities emphasize the relationship between the building and the sea.

The functional program of the museum is highly constrained and the brief stipulated that the 80,000 m² of construction be complete before the end of 2015. Working with Cox Architecture, Arup devised a structure of tilted planar portal frames that enable 3-D curvilinear forms and spaces to be created through parametric variation of the system. The portal frames are stabilized by exposed diagonal bracing that also helps to support the large building cantilevers and provide architectural expression.

It was recognized at the start of the project that a parametric design process was necessary to efficiently analyze and document the buildings while still

Figure 9.2   Structure of a hall.

providing freedom for variation within the design. Cox and Arup collaborated to construct the parametric model using Rhino and its parametric design plug-in, Grasshopper. The architect set up and controlled the components that generated the overall shape of the building and the centerlines of the portal frames, while the engineer dictated the required structural properties and performed topology and section optimization. A script was developed to generate the structural framing of the halls from the architectural input with minimal adjustment.

The topology of the bracing system was developed using an evolutionary structural optimization (ESO) technique originally proposed in the early 1990s by Professors Mike Xie and Grant Steven [1]. ESO is based on the simple concept of slowly removing inefficient, underutilized material from a structure so that the residual topology evolves toward the optimum.

The scripted workflow for the maritime museum is as follows:

1. The parametric centerline model is set up containing bracing in every possible configuration.
2. The parametric model is converted to Arup's in-house finite element (FE) analysis software, GSA.
3. Portal frames and key bracing elements are identified and removed from the optimization.
4. Clashes between bracing elements and architectural elements are detected and resolved.
5. Loads and support conditions are assigned to the FE model.
6. Using a VBA script, the bracing pattern is evolved toward an

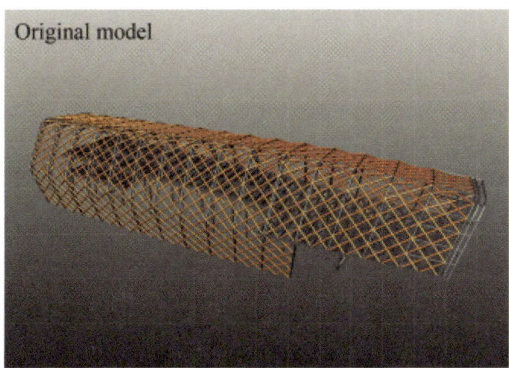
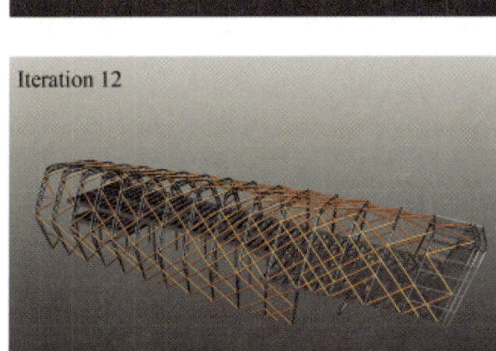
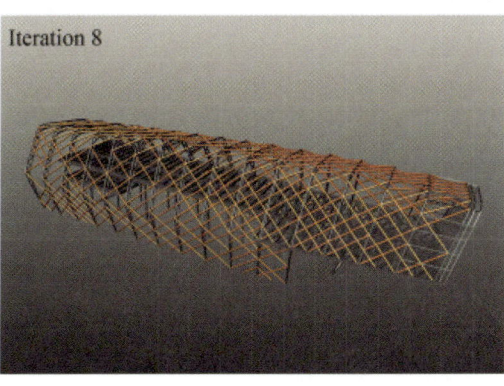

Figure 9.3 Topological optimization.

optimized configuration by iteratively removing the least stressed bracing elements under the envelope of load combinations.

The organic pattern that evolved from the ESO is structurally efficient and complementary to the architectural intent of a fluid design.

Custom Tool Development

Software development is a fundamental component of design computation at Arup. We rely on our ability to extend and customize off-the-shelf software and encapsulated design data to leverage multiple analysis platforms.

DesignLink SDK

The DesignLink SDK is an open-source, .NET software development toolkit developed in-house by Arup. It provides a data structure similar to IFC suitable for storing interdisciplinary building-information models tailored to our specific needs. It also lends interoperability functionality to exchange data between a wide variety of different formats and software APIs. DesignLink facilitates internal software development by providing a common codebase that individual developers can leverage, making it easier to integrate different tools and to share and reuse code between different projects.

SALAMANDER

One of the most significant projects currently utilizing the DesignLink SDK is Structural Analysis Link and Manager for Data Entry and Retrieval (SALAMANDER). SALAMANDER is a plug-in for the modeling software Rhino that allows it to be used as a modeling and preprocessing tool for finite element (FE) analysis models. Rhino is a geometric modeling tool and file format polyglot, making it an ideal companion to engineering analysis packages. SALAMANDER eases the process of transitioning between modeling and analysis environments by linking Rhino geometry to structural analysis data and providing a suite of tools to edit the analysis model. This allows the engineer to build analysis models in Rhino side by side with the architectural model, and to move fluidly between representations as the design is refined without losing the existing analysis data.

SALAMANDER also makes it possible to define structural analysis models entirely parametrically, thus opening the possibility of an algorithmic approach to the design process itself. It provides a set of components that allow it to be driven from the parametric modeling plug-in, Grasshopper. Analysis results can be read back, understood by SALAMANDER, and used in conjunction with an optimization engine to drive the parametric model's input parameters. This allows it to iteratively generate progressively more optimal design arrangements.

Several different optimization tools are used in this context within Arup including Grasshopper's built-in Galapagos genetic algorithm solver. The Arup-developed Gecko component plug-in takes an approach based on satisfying discrete optimality criteria, which is a better approach for structural optimization. SALAMANDER itself also contains an optimization engine custom-built specifically for automated section sizing and a dynamic relaxation solver for geometric and structural form finding.

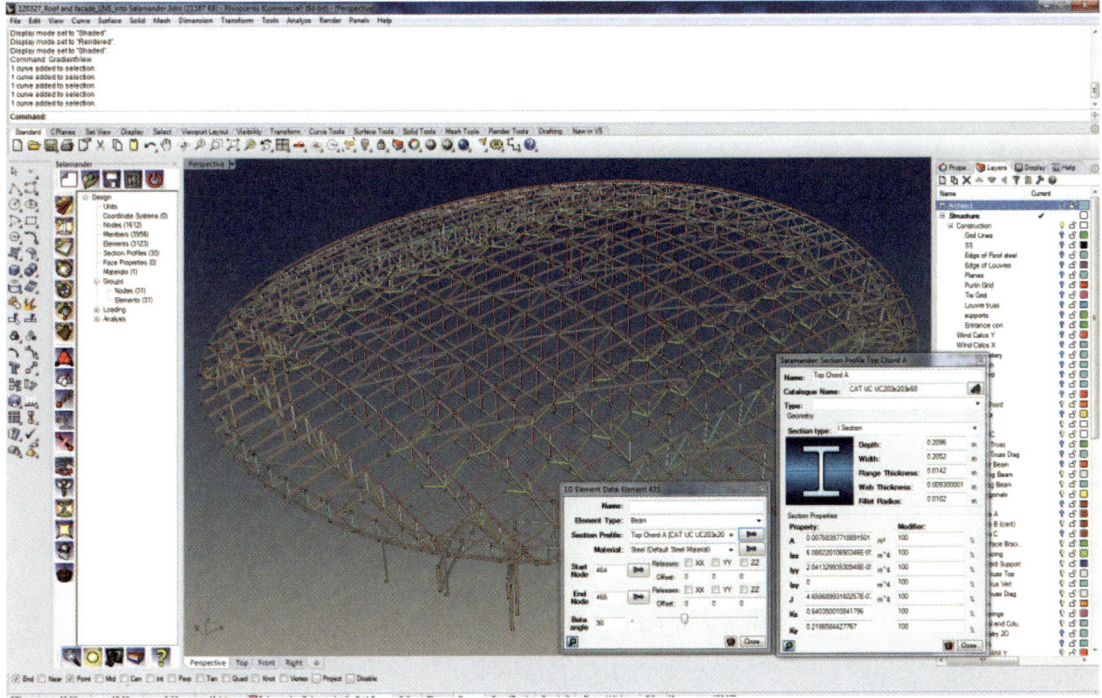

Figure 9.4 The optimized design of the KASC Sports Hall and Athletics Stadium roof structure viewed inside SALAMANDER.

This linked parametric/optimization approach has been used on a number of projects, including the design of the roof of the multipurpose Sports Hall and Athletics Stadium that forms part of King Abdullah Sports City in Saudi Arabia. In this project SALAMANDER and Grasshopper were used to rapidly generate and test different options for the overall roof structure. Once an option had been chosen, the model was refined and used as the basis for geometrical optimization using Galapagos. Preprocessing tasks such as generating the unique wind loads on each element in the structure over the entire domed roof surface were performed in SALAMANDER; the solver was allowed to control several aspects of the geometry including the number, spacing, sizing, and exact arrangement of the long-span roof arches. The optimization process took into account geometric and architectural considerations, such as the clear height afforded at certain locations below the roof, as well as the typical structural criteria of overall steel tonnage, resulting in a final design that was both structurally efficient and tailored to meet the specific project requirements.

Stadium Generator

Stadium generator (StaG) is a toolkit developed for Arup Associates' Sports Architecture team for the rapid parametric generation of spectator seating facilities in sports stadia and other venues. Attaining an optimum design for the seating bowl is of paramount importance in stadium design as it has significant influence on spectator experience and the profitability of the building. The layout of the bowl is also a major driving factor in the design of the rest of the building's geometry. As a result there is significant pressure to develop an optimum design response for the bowl as rapidly as possible.

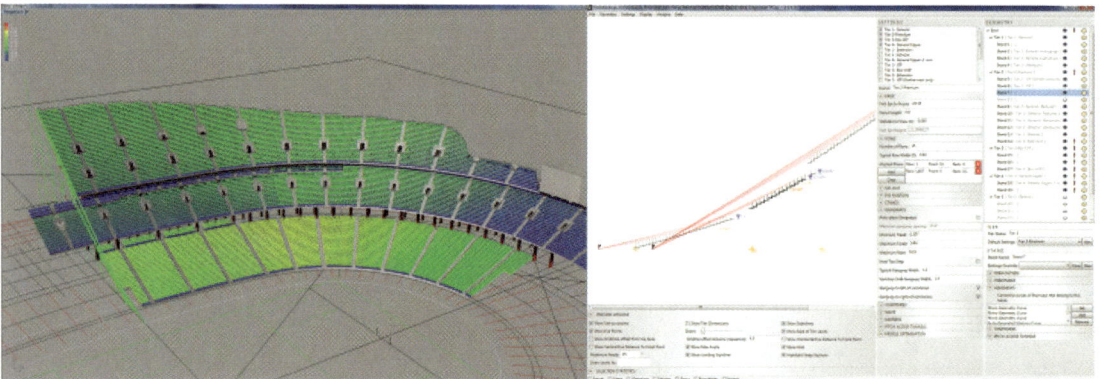

Figure 9.5 A section of Tokyo National Stadium seating bowl undergoing a C-value (spectator view quality measure) analysis.

Optimum seating bowl geometry is driven by a number of factors including codified design requirements and various rules of thumb having to do with ease of construction and spectator experience. Many of these drivers can be mathematically described, making the problem a good fit for a parametric approach. However, the scale and complexity of the analysis make existing general-purpose parametric tools unsuitable and the problem demands a customized solution.

StaG offers a comprehensive range of options for generating stadium seating arrangements from over 60 different key parameters and basic geometrical inputs. A 2-D sectional preview gives instant feedback on the implications of any change in input, and the tool can generate a full 3-D bowl in a matter of seconds, complete with seating, gangway, vomitory, and barrier geometry. The user interface has been developed in close collaboration with the end-user sports architects in order to accommodate their ideal workflow. The program also contains a suite of specialist tools to analyze and visually display detailed information on aspects of the design and the expected spectator experience.

Development of the tool is ongoing; however, it has already been used to great effect on several projects including the winning competition entry for the Tokyo National Stadium developed in collaboration with Zaha Hadid Architects.

Automation

The drive toward a more efficient design process is simple: do more with less. The pressures are threefold. There is the obvious cost pressure to be competitive, but as a firm of designers, we also see the need to execute repetitive tasks more efficiently to free up time to allow creative engagement with our collaborators and the problem at hand. We are also continually pushing the boundaries of what is technically possible and attempting to deliver more intricate designs on a regular basis. This requires us to efficiently execute complex tasks and leverage our design data to perform multiple processes with the same data set (i.e.; modeling, analysis, takeoffs, and documentation).

Much of our computation work is simply automating customized design processes such as design checks, model interrogation, data harvesting and processing, and automated documentation or visualization tools. Most engineering software provides an API for automating its procedures and data I/O that we use to link different analysis procedures together into a complete design platform. In addition, cloud computing, distributed computing frameworks such as HTCondor, the NVIDIA graphics processing unit (GPU) parallel computing platform, and CUDA all have been used successfully at Arup to parallelize computationally intensive design and analysis procedures.

Reference

[1] Xie, Y.M., and Steven, Grant P. "A simple evolutionary procedure for structural optimization," *Computers and Structures*, 49, pp. 885–886, 1993.

Chapter 10

Generic Optimization Algorithms for Building Energy Demand Optimization: Concept 2226, Austria

Lars Junghans

Introduction

The need for computer software in building planning processes to calculate the performance of a project is self-evident nowadays. Building simulation programs are used to calculate the energy demand, the structural load, or even the cost of a building.

Building design and renovation projects are multivariable parameter problems that include a large number of possible combinations of parameter settings. The parametric study often used in planning processes involves changing one parameter while leaving others constant. These studies can miss important interactive effects [1]. One way to find a global optimal solution is to use enumerative search methods where all possible parameter settings are combined with each other. Because of the large number of combinations, however, this optimization process is computationally expensive and would take too much time. A more promising solution is to use an automated building optimization algorithm coupled with a simulation program to find an optimal solution [2].

The term "building optimization" refers to an automated method that uses algorithms to find the optimal combination of parameter settings for building design and renovation. The objective of the method is to find an optimum for the lowest energy demand, cost, or greenhouse gas emission. When building design parameters of the building envelope, the building automation as well the HVAC system can be included in the optimization process. The term automated building optimization indicates that the building optimization algorithm provides optimal solutions without extensive user interaction. The user still needs to define the problem and needs to provide necessary material data. A typical future task of automated building optimization algorithms would be to define the optimal properties of a climate-responsive building façade. In this task, the optimal combination of window to wall ratio, glazing, insulation, and shading geometry will be found to reduce the energy demand for heating, cooling, and artificial lighting. Automated building optimization will even be able to define the shape of a building within a given building program.

In the building sector the goal of a building optimization algorithm should be:

1. Reduction of computation time
2. Robustness of the results
3. User-friendliness of the application of the algorithm.

Reduction of the computation time is especially important for automated building optimization algorithms because they are using time-consuming thermal dynamic simulation software. A calculation time of several hours, like in a multiparameter optimization problem, can be critical in a planning process, especially in the early design stages.

The robustness and reliability of the calculation results in a recommendation of the global optimal solution that is especially important for the user. Some of the current available optimization algorithms are not adjusted for the special needs in building optimization problems, which are different than the needs in other scientific optimization problems with a much larger problem space.

The user-friendliness is important for the use of a building optimization algorithm in planning projects where no expert knowledge is available. Some currently available building optimization algorithms need expert knowledge in the use of the algorithm and its connection to the simulation software tool. This often not existing expert knowledge is a reason for the currently limited use of building optimization software in practicing planning offices.

In this chapter, strategies are described as the elements in a search space. Each strategy has a combination of parameter settings. To define the performance of each strategy a thermal building simulation is necessary. In general, optimization approaches taken toward achieving the objective described above can be classified as discrete and continuous parameter optimization methods.

Discrete parameter methods are mostly used for building optimization. For example, a finite number of available construction types and thicknesses are available when adding insulation to a wall. In contrast, continuous parameter methods do not use fixed numbers for the parameter setting for building shape or dimension such as the window-to-wall ratio, building orientation, or compactness.

Continuous parameter methods based on numerical optimization were studied as early as the 1990s [3]. Although researchers found that numerical building optimization algorithms based on simulations have nonsmooth functions and can fail to find the optimum solution [4], several optimization methods using continuous parameters have been successfully developed for building shape and dimension optimization. These methods include the simplex method [5], the pattern search algorithm [6], the harmony search algorithm [7], the multidirectional search algorithm [8], and the simulated annealing algorithm. However, even given their theoretical success in finding optimal building shapes, these methods are limited because building optimization projects have a combination of discrete and continuous parameters and are not useful for these studies. Instead, optimization methods using discrete parameters, like the genetic algorithm, particle swarm, and sequential search methods, are more suitable.

Optimization Methods: Probabilistic Optimization Methods

1. *Genetic algorithm.* The GA is a probabilistic search technique for solving complicated problems using evolutionary principles to find optimal solutions [9]. It searches for an optimal solution to a multiparameter problem by simulating the natural selection over generations. A potential combination of parameter settings is described as genes on a chromosome. A population is created in every generation where the performance of each population is expressed in a fitness factor. Members of the population

are paired up to create a new generation. The selection for the parents is selected randomly, where members with a better fitness factor have a better chance to be selected.

As with organisms in nature, a crossover of the chromosomes takes place to define the property of the genes of the children. In some GAs, a mutation process takes place to refresh the gene pool. The process of producing new generations is repeated until an adequate optimal solution is found. The GA is seen to be a robust search technique to avoid local minima. However, with each iteration or generation of the algorithm, a different path toward an optimal solution occurs and the end result may also be different [10].

An advantage of this algorithm is that it avoids generating local optimal solutions, which is a problem with the numerical optimization algorithm. The GA has many practical uses in building science, including the optimization of structural systems, building shape [11,12], and building HVAC [13]. However, experiments have illustrated that this algorithm does not always result in good solutions. Currently, the optimization process must be repeated several times to prove that the recommended solution is the global optimum. This repeating simulation process has the disadvantage of extending the overall calculation time of the optimization process. This is the reason why the GA has not found its way into practicing planning offices.

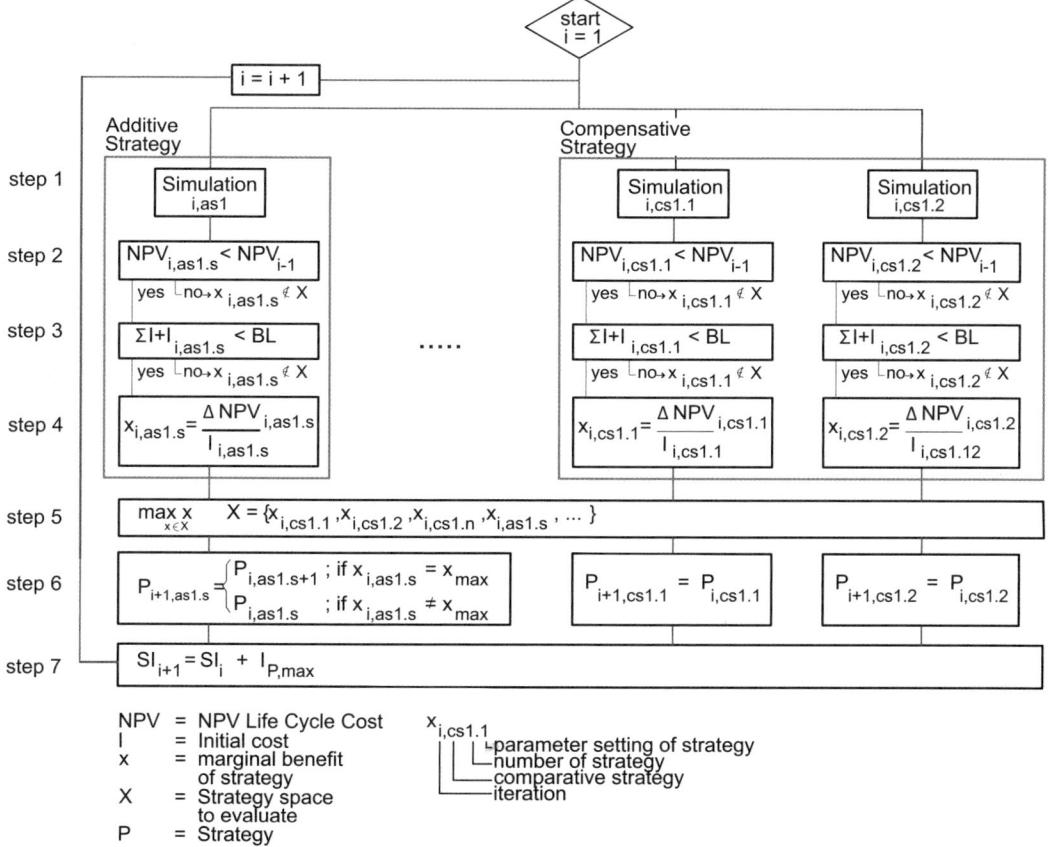

Figure 10.1 Visual explanation of the equimarginal optimization algorithm. It is divided between additive strategies like adding insulation and compensative strategies like the type of glazing.

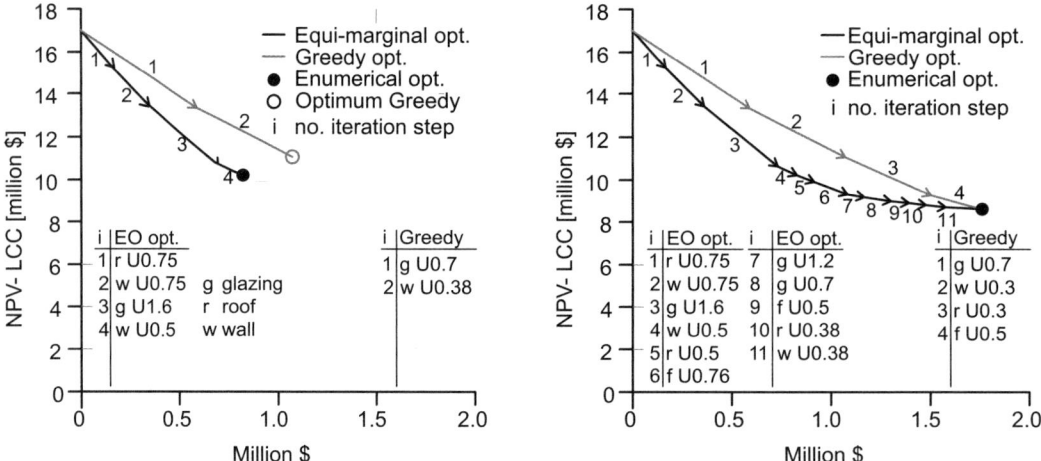

Figure 10.2 This figure illustrates the outcome of the equimarginal optimization algorithm. It shows recommended renovation strategies for a dormitory building in Ann Arbor, Michigan, under different budget limits. The recommended strategies are ranked according to their efficiency in reducing the life-cycle cost. The diminishing effect of return is also visible for the user of the sequential search algorithm.

2. *Multiobjective genetic optimization.* The GA can be used for multi-objective optimization problems. Rather than only having one fitness criterion like the energy demand or the life-cycle cost, a multiobjective optimization algorithm will provide optimal solutions for two or more fitness criteria. The multiobjective GA normally provides pareto-optimal solutions. Researchers have used the multiobjective GA to find optimal solutions for the life-cycle cost and the life-cycle of greenhouse gas emissions.

3. *Particle swarm optimization method.* The particle swarm optimization method has many similarities to the GA and also proceeds by probabilistic parameter settings [14,15]. However, unlike the GA, the particle swarm method is based on the social behavior of birds or schools of fish rather than on evolutionary principles like mutation and crossover. In each iteration step, parameter settings are changed randomly, and strategies, or swarm particles, in the search space are compared to each other. Changes to parameter settings of the most successful particle are adopted by the other particles in the search space. The process is repeated as long as necessary to find the optimum. Although this method has been used successfully for building optimization, the method has been found to be more computationally intensive than the GA [10].

Optimization Methods: Sequential Search Algorithms

1. *Sequential search algorithms.* A sequential search algorithm is a top-down optimization method that iteratively improves the building performance. Unlike other optimization algorithms, sequential search algorithms are not based on randomly defined parameter settings. In each iteration step, the most effective solution is found by comparing the results of previously defined strategies and parameter settings. The process is repeated until an optimal solution is found. A vector path is found from the initial design of the building to the optimal combination of parameter settings. The optimal strategy in the search space is found according to its marginal benefit and recorded in each iteration step. Because of this sequential approach,

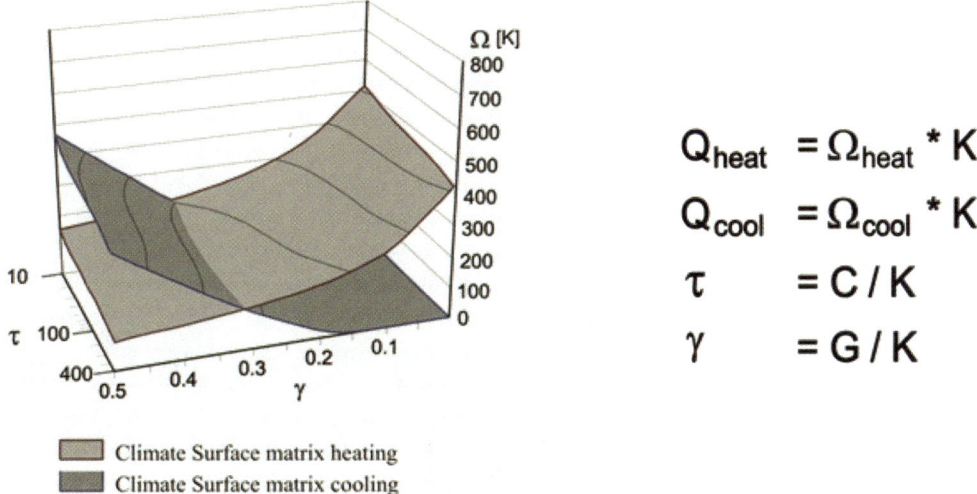

Figure 10.3 The climate surface matrix for fast heating and cooling energy demand calculation.

this algorithm has a significant advantage over optimization algorithms that provide only the optimal solution in that it ranks the recommended strategies and also provides the marginal benefit.

A difference in the sequential optimization process, or greedy search, used for building simulation compared to the problems in microeconomics or mathematics is that the starting values, such as energy demand or life-cycle cost, will change dynamically in each iteration step in relation to reductions in operation energy demand in each successive step. Only top-down sequential approaches can be used because forecasting the results in future iterations is not possible. Thus, there is a risk that large improving strategies can be overlooked in earlier iteration steps.

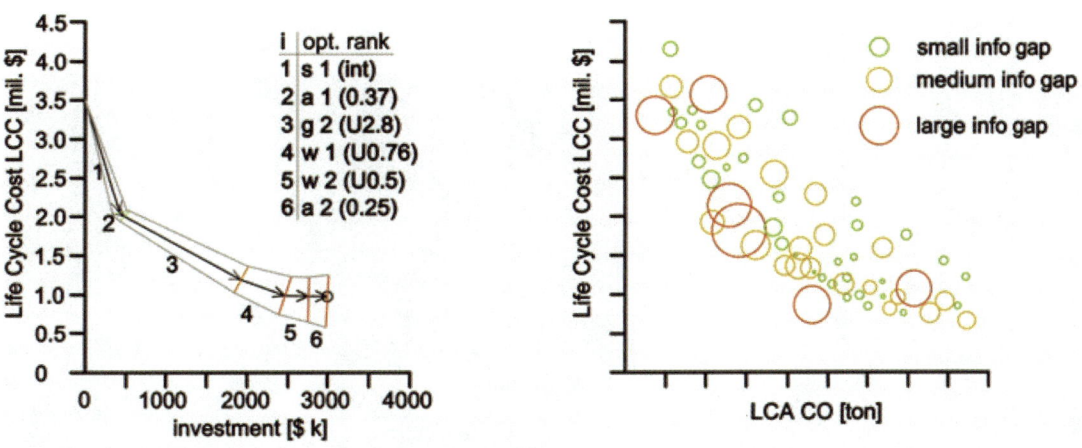

Figure 10.4 These diagrams are illustrating how the uncertainty of the results in building optimization processes can be illustrated. Left: The result of an equimarginal optimization process with the mean-path and the expected deviation or the results. Right: The uncertainty of a multiobjective genetic optimization process.

2. *Equimarginal optimization.* The equimarginal optimization algorithm (EO) is a sequential top-down algorithm that solves the problem of the greedy algorithm and uses concepts from microeconomics [16]. As background, marginal utility (MU) is the benefit or satisfaction from the purchase or consumption of a selected quantity unit of a good or service. The EO is based on the diminishing marginal utility that is the effect when the MU decreases with the increasing quantity of a good or service.

The EO has the advantage compared to other building optimization algorithms that it provides the marginal benefit of the investment for each possible recommended strategy. It also ranks the recommended renovation strategies according to their reduction of energy demand or life-cycle cost.

Figure 10.5   Exterior of Concept 22/26. (Image courtesy of E. Hueber.)

Figure 10.6   Interior of Concept 22/26. (Image courtesy of E. Hueber.)

Architects and decision-makers are thus able to find the balance between the long-term life-cycle cost and the short-term investment. Because the EO is a top-down search method, it has its advantages in building renovation projects, where a nonrenovated building with high energy demand will be improved.

Available Optimization tools

1. *GenOpt.* GenOpt is an optimization tool developed by the Lawrence Berkeley Laboratory in 2001. It is based on the Nelder-Mead pattern search technique, which is a special form of the simplex method. The algorithm needs concave functions in its parameter-setting description. GenOpt can be used as an optimization platform for simulation software tools like EnergyPlus or TRNSYS. Researchers have used GenOpt for building and system optimization where concave functions of the parameter settings are available. So far, expert knowledge is necessary to integrate GenOpt in simulation software.

2. *BEopt.* BEopt is based on a sequential search algorithm and is a version of the greedy search technique. It uses the DOE 2 and TRNSYS as a simulation environment. For each iteration, the most cost-effective strategy is chosen based on the reduction of life-cycle cost. As in most forms of sequential search algorithms or the greedy algorithm, the chosen strategy is then removed from the parameter search space for future evaluation. The developer of BEopt modified the algorithm to overcome the problem of overlooking large improvement strategies and negative interactions between strategies. BEopt keeps track of solutions in previous iterations and compares them with the current solution. Interactions between parameter settings must be identified by the user before using the algorithm so technical expertise is necessary [10].

3. *Genetic algorithm.* A number of free available optimization tools are available that are based on the GA. These tools are mostly based on computer language codes like Java, C++, or Python. The MATLAB program also has a number of GAs in its library. All these tools require expert knowledge in the use of the optimization algorithm. Additionally, computer coding is necessary for their use in building optimization.

Future Developments in Generic Building Optimization

1. *Improvement of usability in the architectural praxis.* Current optimization algorithms need expert knowledge for their use in building optimization. Future building optimization algorithms should be adapted to the needs of building optimization. A goal must be to develop optimization algorithms that can be used by architects and planners without expert knowledge in optimization theory and computer science. The robustness of the outcome of a building optimization process must be improved so that the user can rely on the outcome of the process.

2. *Improvement of the calculation time.* Large energy reductions in the built environment can be achieved in early design stages in the planning process. When a building is designed or renovated, it is often not clear which part of the building envelope or technical system is most effective to renew, improve, or replace. To meet these requirements, the algorithm used in the early design stages must feature a short calculation time to meet economical aspects. The following calculation methods currently

exist and will be discussed in the following sections: static calculations, dynamic calculations, and climate surface calculations.

Static calculation methods that appear in national building codes are based on simple equations with a limited number of values representing the specific climate. An advantage of these methods is that they do not have a long calculation time. However, these methods do not consider the time correlation of outside air temperature, solar radiation, internal thermal mass, and internal heat gains. Therefore, the imprecise results of the static calculation cannot be used for building optimization.

Dynamic calculation methods using dynamic simulation software programs such as DOE2, EnergyPlus, and Trnsys have been introduced into the scientific community. The programs calculate the energy demand in short simulation time intervals of an hour or shorter based on hourly weather data of a reference year. Unlike static calculation methods, advantages of these dynamic methods are that they consider time-correlation, are more precise, and consider different qualities of the built environment. However, the programs are too time-consuming because they are based on a large number of thermal dynamic calculation processes in each iteration step.

Burmeister and Keller [16] introduced a concept called the climate surfaces. Dynamic presimulations are necessary to generate the data matrix for the climate surface of a specific building type. This fast-calculating method can overcome the disadvantage of the computationally expensive dynamic simulation tools. However, the climate surface method needs to be extended with factors like natural ventilation, daylight, shading, and psychrometrics.

3. *Clarification of uncertainties in building optimization.* The GA and EO derive their usefulness in building optimization through the use of thermal dynamic simulation tools for energy demand predictions. These tools are based on a vast set of climate data and can thus produce accurate predictions, which is critical for sustainable and economic building evaluations. However, life-cycle assessments for greenhouse gas emissions and costs in building science commonly suffer from significant uncertainty due to lack of information. Uncertainties occur as a result of differences between the physical reality and the simulation calculation model [17], cost definitions [18], and definition of the specific environmental impacts of building material and systems. Because of these uncertainties, the proposed optimal solution given by computer-based building optimization can create unreliable results [19].

For building planners, designers, and project decision-makers, it would be very helpful to be able to quantify the uncertainties of the results in a building optimization process. Depending on the assessment criteria, an option with a smaller predicted return and a smaller risk might be preferred to an option with a larger predicted return and a larger uncertainty or risk. When information is severely uncertain, a decision-maker may want to make a decision that will yield a reasonably satisfactory result over a large range of realizations of the uncertain parameters.

Built Example

Concept 2226 is a highly innovative office building located in the cold climate of Austria. The building is named 2226 because it maintains the comfortable temperature range of 22–26 degrees Celsius without any active heating, cooling, and ventilation systems. The total energy demand for heating and cooling is 0 kWh/m²a. The project was designed by the Baumschlager Eberle Architectural office. The 2226 building has a gross floor area of 3,201 m² and was constructed in 2013. A combination of several unique design features, spatial organization, and sophisticated software allow the optimal management of energy flows. The building is equipped with automated window openers. The natural ventilation is controlled according to the occupants' demand. A building optimization algorithm is used to find optimal building automation set-points for the temperature, CO2 concentration, and absolute humidity.

Conclusion

Automated building optimization algorithms and systems are becoming more and more useful in architectural design processes. Great energy use reductions, cost reductions, and greenhouse gas reductions are expected by the use of these generic software tools. By the integration of faster calculating simulation tools, optimization processes will be applied by architects in early design stages.

In the near future, architects will be able to design buildings and building elements with a higher confidence in optimal solutions than in the past. Results will be presented as continuous values, including the mean value, expected value, and the deviation. An improved risk management will be possible. It will be up to the architect or decision-maker to decide how far an automated building optimization decision support system will go.

References

[1] Andersen, R., Christensen, C., Barker, G., Horrowitz, S., Courtney, A., Gilver, T., Tupper, K., 2004, "Analysis of Systems Strategies Targeting Near-Term Building America Energy-Performance Goals for New Single-Family Homes" NREL/TP 550 36920.

[2] Wright, J., "Optimization of Building Thermal Design and Control by Multi-Criterion Genetic Algorithm." *Energy and Buildings.* 34,959-972.

[3] Bouchlaghem, N.M., Letherman,T., "Numerical Optimization Applied to the thermal Design of Buildings," *Building and Environment* .Vol 25 2 (1990) 117-124.

[4] Choudhary, R., Malkawi, A., Papalambros, P.Y., Analytic target cascading in simulation-based building design, *Automation in Construction* 14 (2005), 551–568.

[5] Bouchlaghem,N., "Optimising the design of building envelopes for thermal performance," *Automation in Construction* 10 Ž2000. 101—112.

[6] Wetter, M., "GenOpt Generic Optimization Program User Manual, Version 2.0.0," Lawrence Berkeley National Laboratory Report, LBNL 54199, Berkeley CA.

[7] Fesanghary, M., Asadi, S., Geem, Z.W., "Design of low-emission and energy-efficient residential buildings using a multi-objective optimization algorithm," *Buildings and Environment,* 49 (2012) 245–250.

[8] Dennis,J.E., Torcson,V., "Direct Search Methods on Parallel Machines," *SIAM Journal of Optimization.* 1991.

[9] Leu,S.S., Yang,C.H., "GA based multicriteria optimal model for construction scheduling." *Journal of Construction Engineering and Management.* 125 (6) (1999) 420-427.

[10] Tuhus-Dubrow, D., Krarti, K., "Comparative Analysis of Optimization Approaches to Design Building Envelope for Residential Buildings," *ASHRAE Transactions 2009*, 115 554-562.

[11] Wang, W., Rivard, H., Zmeureanu, R., "Floor shape optimization for green building design," *Advanced Engineering Informatics* 2006 20 363-78.

[12] Ourarghi,R., Krarti, M., "Building shape optimization using neutral network and genetic algorithm approach," *ASHRAE Transactions 2006.* 112 484 – 91.

[13] Mossolly, M., Ghali, K., Ghaddar, N., "Optimal control strategy for a multi-zone air conditioning system using a genetic algorithm," *Energy 2009.* 34 58-66.

[14] R. Bornatico,M. Pfeiffer,A. Witzig, L. Guzella, "Optimal sizing of a solar thermal building installation using particle swarm optimization," *Energy.* 41 (2012) 31-37.

[15] Krugman, P., Wells, R., Economics, Chapter 10, p. 259, Worth Publisher New York 2009, ISBN 13.978.0.7167.7158.6.

[16] Burmeister H., Keller B., Climate surface: a quantitative building-specific representation of climates, *Energy and Buildings.* 28 1998 167-177.

[17] Williamson, T.J., 2010, "Predicting building performance: the ethics of computer simulation," *Building research & Information.* 38 (4) 401-410.

[18] Bristow, D., Kennedy, C.A., 2010, "Potential of building-scale alternative energy to alleviate risk from the future price of energy," *Energy Policy.* 38 1885-1894.

[19] Rysanek, A.M., Choudhary, R., 2013, "Optimum building energy retrofit under technical and economic uncertainty," *Energy and Buildings.* 57 324-337.

Chapter 11

Customized Algorithmic Engineering of a Curved Cable-Stayed Façade: The Enzo Ferrari Museum, Modena, Italy

Lucio Blandini, Timo Schmidt, and Werner Sobek

Introduction

Today's demand for special geometries and highly transparent building envelopes calls for innovative engineering solutions. The following article presents the exhibition gallery of the Enzo Ferrari Museum in Modena, Italy, outlining the particular project challenges and showing how cable structures allow for extremely transparent curved façades.

The recently opened museum dedicated to Enzo Ferrari in Modena plays with the duality between the renovated historical building where Ferrari was born in 1898 and a futuristic exhibition gallery designed by Jan Kaplicky (Future Systems, London), shortly before his death. The gallery embraces the masonry building and relates to it, whereas its sculptural form is clearly inspired by sports car design. From the gallery the view converges through the transparent curved façade to Ferrari's birth house, as if you were

Figure 11.1   The curved roof of the new Enzo Ferrari Museum. (Source: Studio 129.)

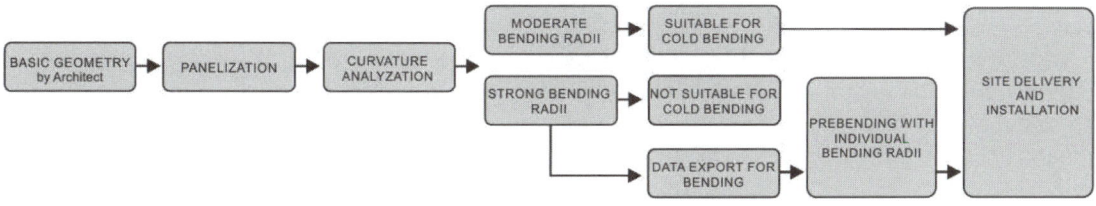

Figure 11.2   Workflow diagram with Grasshopper data for the skylight engineering, manufacturing, and assembly. (Source: Werner Sobek, Stuttgart.)

Figure 11.3 Enzo Ferrari Museum. (Source: Studio 129, Stuttgart.)

looking though an oversized car windshield. The engineering philosophy chosen for the façade of the Enzo Ferrari Museum in Modena was to maintain a relatively simple geometry for the façade panels, which had to be assembled and to be adapted to the different geometrical situations by means of complex customized detailing.

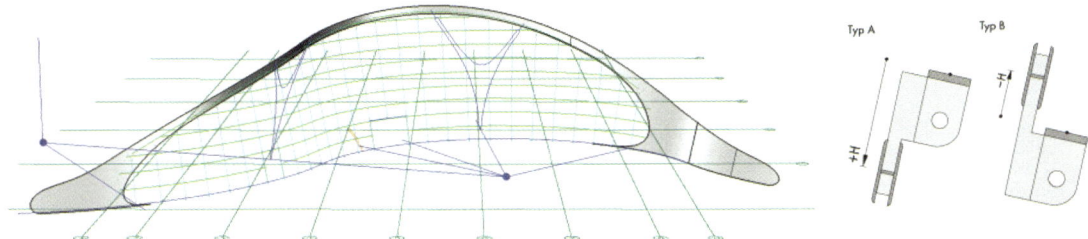

Figure 11.4 Geometrical description of the façade and schematic view of the top details. (Source: Werner Sobek, Stuttgart.)

Geometry

The 11m high cable-stayed glass façade is geometrically defined by two intersecting conical surfaces inclined toward the interior by 12.5°. The sinuous form of the façade was accomplished using straight cables and regular planar glass units. These had to be cut with specific angles to match the conical geometry. Only the upper glass units have a less regular geometry, due to the 3-D curved edge, which is generated from the intersection between the roof surface and the conical façade surfaces. Special details have been developed at the top of the façade, solving parametrically all the different connection situations between the irregular glass units and the supporting steel structure.

The geometric result of this intersection edge is a 3-D curved circular hollow steel girder with a length of 62m. The curved girder has a diameter of 1,000 mm. It constitutes a top-side support for the cables of the façade. The girder was geometrically optimized, resulting in the definition of 13

Figure 11.5 View of the construction site during the erection of the 3-D curved girder. (Source: Studio 129.)

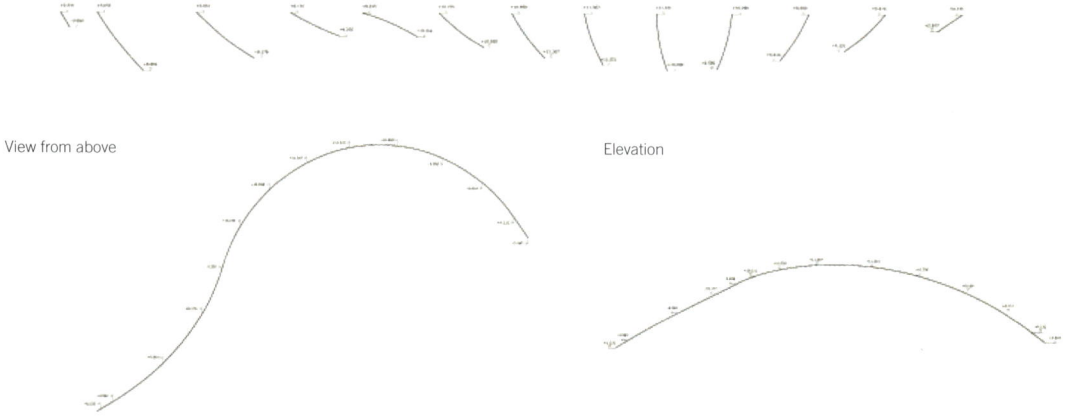

Figure 11.6 Geometrical segmentation of the 3-D curved girder. (Source: Werner Sobek, Stuttgart.)

segments that could be curved only in one direction. The optimization process was controlled by setting tolerance limits between the original geometry and the segmented one. The wall thickness of the steel elements varies between 20 and 40 mm in order to match the different loading conditions. The steel elements were fully welded on site.

Cable-Stayed Glass Façade

A set of 32-mm vertical stainless steel cables supports the flat insulated glass units, made of a 10 -m fully tempered glass pane outside and two 6mm heat-strengthened glass panes laminated with a SentryGlas®Plus interlayer inside. The cavity filled with argon and the solar control coating on face 2 allow for a $U_g = 1.0$ W/m².

Figure 11.7 Enzo Ferrari museum during the opening. (Source: Werner Sobek, Stuttgart)

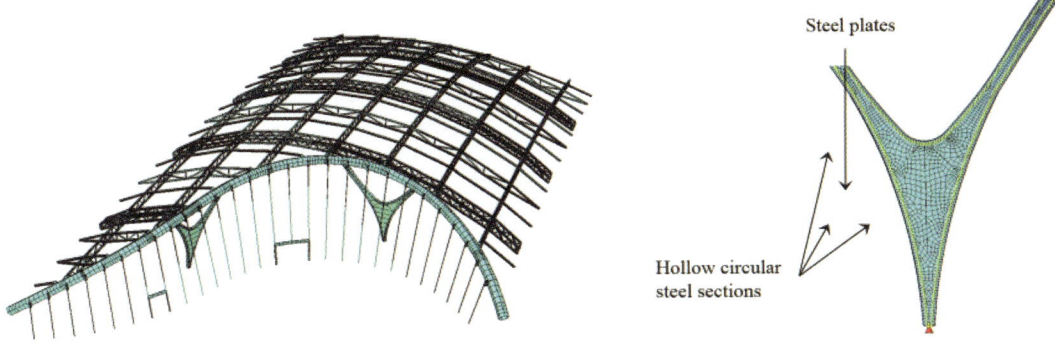

Figure 11.8 Structural model of the roof and façade and special structural model of one Y-shaped column supporting the façade top girder. (Source: Werner Sobek, Stuttgart.)

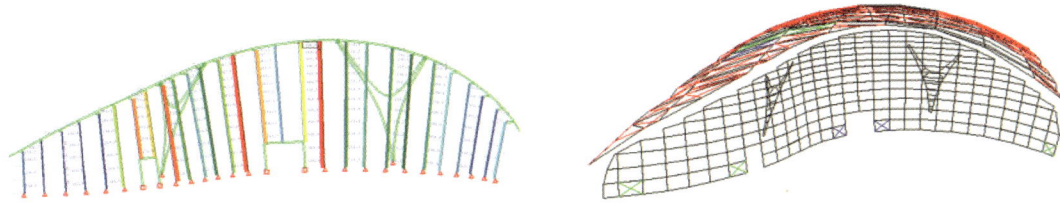

Figure 11.9 Façade structural model: prestress forces and warping check. (Source: Werner Sobek, Stuttgart.)

Figure 11.10 View of the entrance portal (Source: Studio 129) and schematic section. (Source: Werner Sobek, Stuttgart.)

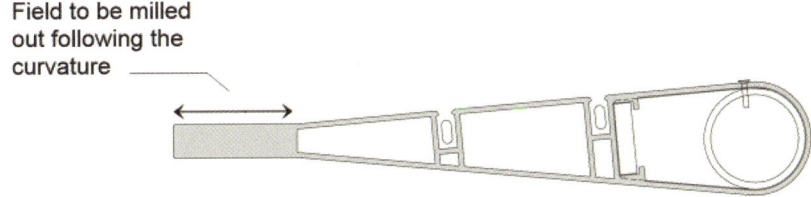

Figure 11.11 Sun shading element—schematic section. (Source: Werner Sobek, Stuttgart.)

Figure 11.12 View of the façade with sun shading elements. (Source: Studio 129.)

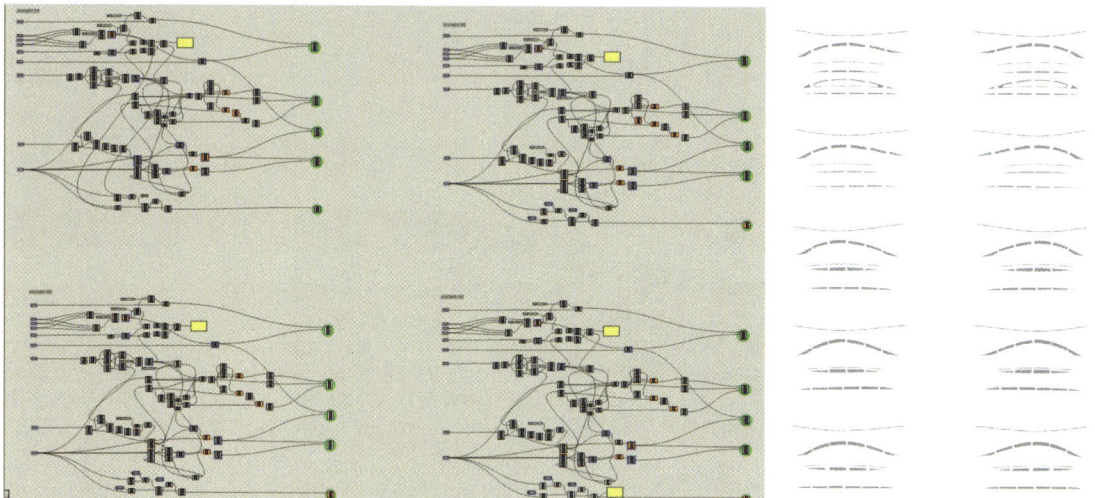

Figure 11.13 Grasshopper data for the skylights and output geometries. (Source: Werner Sobek, Stuttgart.)

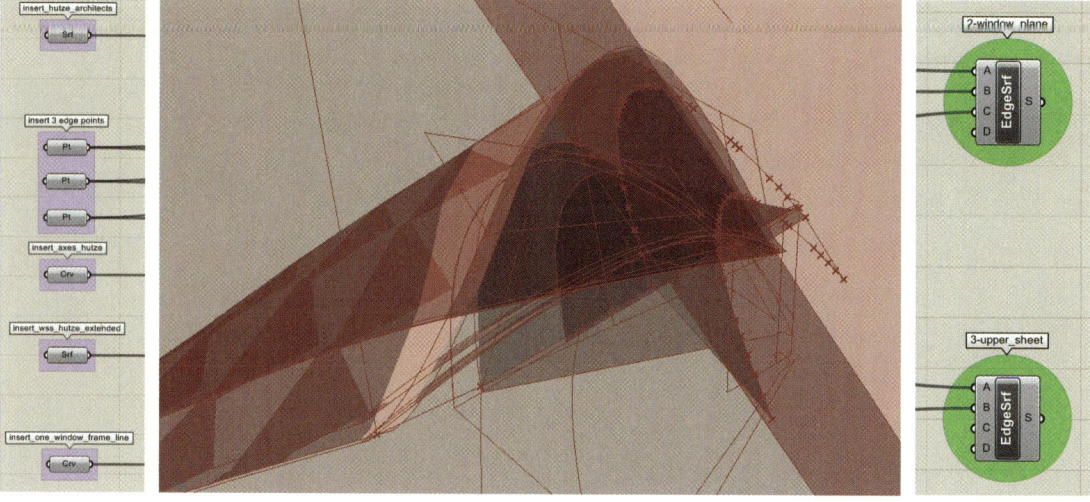

Figure 11.14 Grasshopper input, output, and preview window of the parametric skylight file. (Source: Werner Sobek, Stuttgart.)

 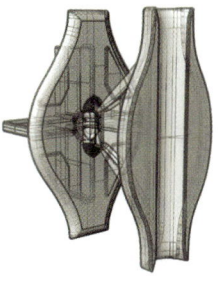 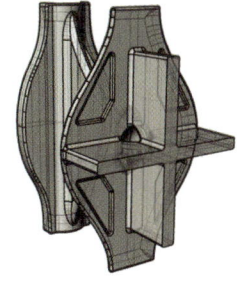

Figure 11.15 a) View of the glass fixing and cable clamp during erection. (Studio 129.)   b) Views of the 3-D model used for casting the piece. (Source Werner Sobek, Stuttgart.)

Figure 11.16   View of the entrance portal façade. (Source: Studio 129.)

The cables not only support the glass units, but also transfer all the loads to the 3-D curved girder mentioned above. The girder is supported by two hinged Y-shaped steel columns and braced by the roof steel structure.

Special attention was paid in controlling the deflections of the whole façade as well as the warping of the most critical insulated glass units by optimizing every single cable pretension force. These vary from 80 to 330 kN.

Structurally speaking, the two entrance steel portals are independent of the façade so that lateral deflections due to wind loads do not hinder the correct functioning of the door-opening mechanisms. A black-coated steel sheet merges façade and door optically according to the architectural idea, whereas joints allow for the arising movements.

Black-coated curved aluminum shading devices further characterize the façade design, recalling a car radiator front and contributing to the reduction of the museum's heat gain. The extruded profiles are not curved themselves; however, the outer part of the profiles was milled out in order to form a curved outer edge. Two heating bands per profile allow for a snow-less element, thus leading to less loadings and therefore less aluminum.

The shading elements as well as the glass units are fixed to the cables by means of a special casted connection element. This connection element was designed specifically for the Ferrari Museum with the aim of reducing the material used to a minimum and to match the specific architectural language. The 3-D model of the pieces was defined precisely in collaboration with a small casting factory specialized in Italian design objects.

Aluminum Roof

The 77m long and 43m wide freeform aluminum roof has to withstand wind, rain and snow while still being able to freely deform due to temperature variations and the absence of any expansion allows for controlled movements and a parametric approach were the keys to address this task.

The outer skin is made of a system of coated extruded aluminum profiles, developed by Pinical. It adopts solutions typically used in the ship building industry. The profiles are joined through a male-connected to a steel secondary structure through adjustable screwed connections. The profile curvature was mainly achieved by bending the profiles on site along their longitudinal axis. The width suitable for site-warping without visually detectable tessellation. Double curved roof elements showing small bending radii were prebent in longitudinal direction and additionally warped on site. This double curved skins without prior thinning the material through plastic deformation.

The analysis of the Gaussian curvature on a 3-D surface model helped identify the critical zones around the skylights and at the backside of the roof. The roof was designed to withstand high temperatures without any expansion joints through a system of movable and adaptable supports. It is fully floating but fixed at certain points along two axes, all directions. The roof temperature was assumed to reach up to 80°C on a sunny summer day and -20°C in winter. Taking a temperature difference of 100 K into consideration, the maximum thermal edges. The node details therefore have to allow the roof surface to slide up to 6 cm in longitudinal and transversal direction. It was indispensable during installation to constantly measure the temperature according to the surface temperature. Beside the sliding ability, the node detail can adapt to the different roof inclinations and is adjustable in height. This allowed utilizing only one typical detail between the segmented support surface (steel structure, insulation and waterproofing layer) and the smooth outer surface. In order to trace the roof curvature, 2.500 laser-cut steel plates were bended into L-shaped profiles to generate a unidirectional substructure. Around 62.500 boreholes were positioned using parametric points. This way more than 95% of the roof surface could be quite easily engineered.

The most complex scripting work was carried to manufacture the 400 laser-cut metal sheets of the ten skylights. Basic geometrical information about the double curved stainless steel sheets and the related edge lines was extracted from the architectural 3-D model. A curvature analysis followed to distinguish elements with moderate bending radii from the elements with strong bending radii. In the first ones the double curved surfaces could be obtained by cold bending the metal sheets; in the second ones pre-bending was necessary, and the relative information was obtained out of the scripts. Also the position of the bolts welded on the rear side of the metal sheets was defined through scripting. This way 25,200 bolt positions were defined.

Conclusion

The façade engineering of the Enzo Ferrari Museum was a considerable challenge due to the required transparency and the geometrical complexity. Such complex freeform skins call for a change in the planning process from 2-D drawings to 3-D digital models. Today, standard planning tools support a quick architectural design of highly complex surfaces. However, the detailing, production, and erection of such envelopes remain a very complex and challenging task. In the Enzo Ferrari Museum the influences from the automotive industry are not only visible in certain design elements, but also in the approach to detailing and in the required precision. The achieved building quality was made possible not only by an innovative façade engineering approach, but also by the close cooperation between the client, the architects, the engineers, and—last but not least—the contractors.

Acknowledgments

The authors wish to thank the Foundation Casa Natale Enzo Ferrari for having made this museum possible. Special thanks are due to the architect Andrea Morgante (Shiro Studio), to the structural engineer Fabio Camorani (Politenica), and all the contractors involved (CdC, Teleya, Pinical, etc.) for a very good collaboration.

References

[1] Future Systems, Shiro Studio: Museo Casa Enzo Ferrari, Modena, Milano, Electa architettura, (2012).

[2] Blandini, Lucio, Schmidt, Timo, Winterstetter, Thomas, Sobek, Werner: The Enzo Ferrari Museum, Modena—Engineering a Free Form Skin. Proceedings of "Advanced Building Skins," Graz, (2012), S. 27–28.

[3] Blandini, Lucio, Schmidt, Timo, Winterstetter, Thomas, Sobek Werner: Das Enzo Ferrari Museum—eine Fusion zweier Designwelten. Glasbau 2013, Editors Bernhard Weller & Silke Tasche. Ernst & Sohn, Berlin, (2013), S. 1–10.

Part III

Post-Parametric Automation in Construction

Alfredo Andia and Thomas Spiegelhalter

In Part II, we observed the emerging narratives of automation in architecture and engineering firms. Part III presents several cases of automation from the construction perspective. The assembly of a building is the big bulk of men hours and resources spent in the construction process. On average, in the developed world for every $1 that it is spent in design, $20 are spent in construction. Thus, the automation narratives that lead to savings from the construction perspective will have a significant impact in the final cost of construction.

Building manufacturing and automated construction has been a dream that has haunted the building industry since the 19th century as many other industries moved from craft-based production to mass-manufacturing in the 20th century and to highly digitally automated manufacturing more recently.

We begin this section by looking at significant digital automation manufacturing processes in the car manufacturing industry. Chapter 12 outlines some of the latest parametric algorithmically automated optimization tools for entire factory planning and operation processes in the automotive and transportation industry. The case study features the interoperable real-time plant simulation software utilized in the example of the Volkswagen Group's use of robotic software platforms and automation. Real life-cycle requirements can be simulated in various situations, such as ergonomic human-machine interactions, maintenance intervals and shift patterns, or infrastructural resource input and output flows. Once a factory is built and in operation, the digital life-cycle metamodel enables users to run optimization experiments and what-if scenarios without disturbing an existing production system of the entire plant. The case study also explains the latest developments for learning algorithms based on neural networks, and wasp swarm optimization of logistic systems and automation in the fields of engineering, manufacturing, infrastructure, city, and building automation. In particular it describes how machine learning and adapting mechanisms are used for black box modeling and optimizations of process automation in self-learning factories and product distribution systems with other factories.

Despite the automation advances in other industries, most of the construction industry is still very much dependent on craft-based assembly. There have been several versions throughout history that have attempted to move construction into manufacturing settings. A first version of industrialized construction emerged in the 19th century within the work of companies such Eiffel & Cie, in prefabricated company towns, in early skyscrapers, in Sears Roebuck Co. prefab houses, and in iron and glass structures such as Crystal Palace. A second version of manufacturing construction appeared in massive social housing programs using reinforced concrete prefabrication in the 1950s and 1960s, but there was no evidence these methods were cheaper or faster than traditional construction techniques.

An initial version of digital automation in construction emerged in the research labs of the five largest design-construction companies in Japan in the 1990s, as mentioned in Chapter 1. Most of these earlier Japanese automation efforts were about taking robots and other automated apparatuses onto the job site. Instead in the West and China we see an important trend to move construction manufacturing off-site, making the construction location a place in which different components get assembled.

We observe that there are two types of automation narratives in the design to manufacturing process. We see contractors moving into manufactured prefabrication and architects and engineers pushing for custom fabrication.

The Broad Group (Chapter 13) and Sekisui Heim (Chapter 14) case studies are two cases of major construction prefabricators who are using highly refined manufacturing and assembly systems to significantly reduce costs and environmental impact. Both companies also focus on reducing the carbon footprint with notable energy, water, and life-cycle resource savings in their operation.

In Chapter 13, titled "Prefabricating a More Sustainable Building and Assembling It in 15 Days: Broad Group, China," Alfredo Andia presents one of the most radical cases how the construction industry is profoundly transforming its manufacturing process. Broad Group has created a system which allows 93% of the building to be manufactured in a factory. They have built more than 25 buildings, including a five-story building that was built in 1 day and a 30-story hotel in 15 days. However, the main theme for Broad Group was not to do a fast prefab building but to do a significantly sustainable building. Their approach was very different from designers who use building simulation modeling to predict the performance of the building. They felt they had to reengineer the whole building. They altered the whole load-bearing structure so the building weighs less, uses less materials, uses less energy, generates and analyzes its own real-time data, and is modular so it can be quickly assembled on the site with no waste, no water use, and a significant reduction of transportation to the site.

In Chapter 14, Jun Furuse, Masayuki Katano, and Thomas Spiegelhalter present the workflow and methods of Sekisui Heim Automated Fabrication and Assembly, Minatoku, Tokyo, Japan which is a major construction prefabricator who is using highly refined automated manufacturing and assembly systems to significantly reduce costs and environmental impact. Sekisui also further developed a computer-based enterprise resource planning (ERP) system for controlling the production and logistics flow from the 1980s towards the Heim Automated Parts Pickup System (HAPPS). The system automatically translates design parameters and plans from architects, engineers, and clients directly into parametric-algorithmically processed production plans, bills of materials (BOMs) and data needed to operate the fully automated production. Sekisui builds and sells approximately 13,000 Heim-Unit houses annually, including their Zero-Utility Cost House with solar energy generation systems. All the units are prefabricated in the factory to 80% completion, which minimizes significantly the assembly process.

In contrast to major contractors such as Broad Group and Sekisui Heim, architects and engineers prefer to promote custom fabrication processes instead of manufactured prefabrication largely because of their design concern. We present two cases of construction automation and prefabrication from the designer's perspective.

Alfredo Andia in Chapter 15 elaborates the case of "NBBJ: Customized Prefabrication in Two Hospitals." The main driver of prefabrication in the two hospitals designed by the firm NBBJ was not manufacturing construction but the aesthetics and functional aspects that require a high level of customization for sophisticated health-care clients today. Both hospitals had very unique in-patient rooms, bathrooms, casework layouts, and very complex building systems. The team looked for prefab modular elements in the market but could not find elements that fit their standards that could help deliver better care. Customized fabrication was used to increase the precision and coordination of the whole delivery process.

Chapter 16, titled "Robotic Fabrication: ICD/ITKE Research Pavilion 2012," by Achim Menges and Jan Knippers, from the Institute for Computational Design (ICD) and the Institute of Building Structures and Structural Design (ITKE) at the University of Stuttgart in Germany, portray their research pavilion project in custom fabrication as a way to increase design performance and quality, which is an emerging trend among designers. The pavilion was entirely robotically fabricated from carbon and glass fiber composites. It has only a shell thickness of 4 mm of composite laminate while spanning 8 m. The architects were inspired by the biological model of the exoskeleton of the lobster (*Homarus americanus*) for its local material differentiation, which eventually served as the role model of the entire design-to-robot-fabrication process. The abstracted morphological principles of locally adapted fiber orientation constitute the basis for the computational form generation and 6-axis robot manufacturing process with fiber spool and resin bath layers on prefabricated wood and steel frames. The architects associate their computational design and robotic production to a biomimetic design methodology. They argue that the concurrent integration of the biomimetic principles of the lobster's cuticle and the logics of the newly developed robotic carbon and glass fiber filament winding within the computational design process enable a high level of structural performance and novel tectonic opportunities for architecture.

The design granularity, the sharpness, and the texture of the spaces we inhabit can acquire another level of finesse with customized fabrication and in particular with robotic manufacturing, as demonstrated in the last case study.

Going back to the first chapter of this section, the robotic technology we find in the car manufacturing industry is still very inflexible, expensive, and needs high levels of expertise to operate. Robotic technology is still in an age very similar to when each computer company or cell phone provider had their own operating system and was not connected to the Internet.

However, the robotics technology is about to rapidly accelerate its pace of development.

There are two items that have hindered the explosion of robotics. The first one is a robust operating system across platforms that can propel robot software development; just as iOS and Android did for the explosion of applications in the smartphone platform. Google's recent acquisition of a large number of robotic companies seems to be geared toward consolidating a universal operating system for robots.

The second issue that has stalled robotics is that the large computation power robots need has been traditionally onboard. This is a largely expensive architecture, that is battery-demanding, cooling-intensive, and creates a space problem. The emerging concept of Cloud robotics will allow robot apparatuses to be connected to the Internet, which permits the migration of their intensive need for computation power to the grid on demand. Robots connected to the Cloud promote collective robot learning from other robots' experiences and an instantaneous open access to new algorithms—all aspects that will significantly reduce the cost of computing, weight, battery, cooling, and space requirements of robots on the job site.

A robust robotic operating system and an explosive number of applications in the Cloud will significantly lower the cost and expand the design of robotic apparatuses well beyond the fixed robotic arms we see in manufacturing lines today, eventually transforming the craft-based construction assembly processes and construction equipment that remain relatively unchanged since they consolidated during the industrial revolution.

Chapter 12

Siemens Digital (Self-Learning) Factories and Automation: Automated System Optimization via Genetic Algorithms

Thomas Spiegelhalter

Today many sophisticated software applications are available to design and engineer infrastructures, buildings, and industries. In this chapter, we outline the latest parametric-algorithmic and automated 3-D/4-D optimization tools and how they are utilized to plan and operate entire factories and production lines in the automotive and transportation industry. We survey Siemens AG, a German multinational engineering and electronics conglomerate headquartered in Munich and Berlin, Germany. Worldwide, it is organized into five main divisions: Industry & Factories, Energy, Healthcare, Infrastructure & Cities, and Siemens Financial Services (SFS). Due to Siemens broad range of existing activities, we will focus on the following objectives only:

1. Digital factory design and real-time operation with PLM software
2. Case study of the Volkswagen Group's use of Tecnomatix robotic simulation
3. Genetic algorithms (GAs), neural networks, and wasp swarm optimization of logistic systems and automation

We will conclude the first two sections by discussing the research significance of new GAs and automated swarm intelligence to understand practical implementation in the fields of engineering, manufacturing, automation, and logistics.

Traditional 2-D Factory Design Processes Are Prone to Error

Numerous disciplines participate in the design and installation of a typical factory, including groups responsible for basic architecture, industrial engineering, and logistical planning. These disciplines often find it difficult to communicate effectively, properly control design revisions, and manage the release process or account for external supply chain activity. Collaborative factory design and management optimization provide diverse teams with a single environment to direct large volumes of facility data.

Manufacturing design and operations are also driven by a set of business imperatives that force companies to implement technologies, processes, and practices that not only enable them to compete and profit, but in some cases survive. In order to hit the market window with the right sustainable product at the right time, companies must have manufacturing facilities and operations that are sustainable, flexible, innovative, and agile. They must also have the ability to launch a quality product cost effectively and just-in-time (JIT).

Inefficiency usually begins in the early stages of designing and planning a factory, especially during the scripting of resource and material flow, multiple processing, circulation streams, facility organization, and work cell

Figure 12.1 Parametric production lines and total life-cycle scenarios with SIEMENS-PLM and Tecnomatix Team software of the Daimler Factory in Sindelfingen-Stuttgart, Germany. (Images courtesy of Siemens, Pictures of the Future, fall 2007.)

layout. Often the processes of cross-functional teams fail to communicate and share data effectively. Antiquated tools, such as 2-D drawings, are unable to properly account for the impact of all dynamic factory design-built-operated processes, equipment placement, or material flow throughout the facility. Traditional factory design processes are prone to error. It is inevitable that design teams are too overwhelmed to understand the impact of equipment selection and equipment placement on the factory floor when expressed in 2-D layout environments. It is obvious that manufacturing design with assembly processes require precise, flexible, and adaptive planning of the facility's delivery concepts, storage needs, and transport concepts (Figure 12.1).

Manufacturers like automotive original equipment manufacturers (OEMs), and suppliers want to launch products at a much faster rate while delivering new models to a market that expects innovative products in a timely and cost-effective manner. Planning accuracy and efficiency will increase this approach to help minimize capital investment and maximize return on investment (ROI). An evolving and emerging technology that will enable companies to achieve timely and lucrative product launches is digital manufacturing.

Digital Factory Design and Operation with PLM Software

PLM stands for product life-cycle management and is part of the SIEMENS globally scalable, commercially available software to digitally design, operate, and optimize entire factories and their infrastructures. One of the most utilized software platforms in this portfolio is called Tecnomatix, which is a multidimensional digital manufacturing software engine powered by Teamcenter.

PLM Tecnomatix enables the user to create structured, hierarchical models of complete production facilities, lines, and processes. This is achieved through object-oriented architecture and modeling capabilities. PLM users are assisted in creating and maintaining highly complex systems, including advanced control mechanisms, capsulation, inheritance, and hierarchy. The 3-D engine captures processes in compliance with an object oriented, generic domain model. The generic metamodel is able to simulate entire complex factory engineering and production systems and processes. Simulation includes the architecture of the factory, production processes, warehouse capacities, and equipment utilization. This also includes any logistics, transportation, and military and mining applications in the automotive and transportation industry. The plant simulation's user interface follows Microsoft Windows standards, creating a familiar and easy-to-learn software environment. Simulation models can be created rapidly by using components from application object libraries dedicated to specific factory design and business processes, such as assembly or car body manufacturing. Users can choose from predefined geometries, systems, resources, order lists, operation plans, and control rules. By extending the library with their own objects, users can capture the best-practiced engineering experiences for further simulation studies.

Beyond all this, real life-cycle requirements can be simulated in various situations, such as ergonomic HMIs, maintenance intervals, and shift patterns. The digital metamodel enables users to run experiments and what-if scenarios without disturbing an existing production system or—when used in the planning process—long before the factory or a robot infrastructure is assembled. Extensive analysis tools, statistics, and charts

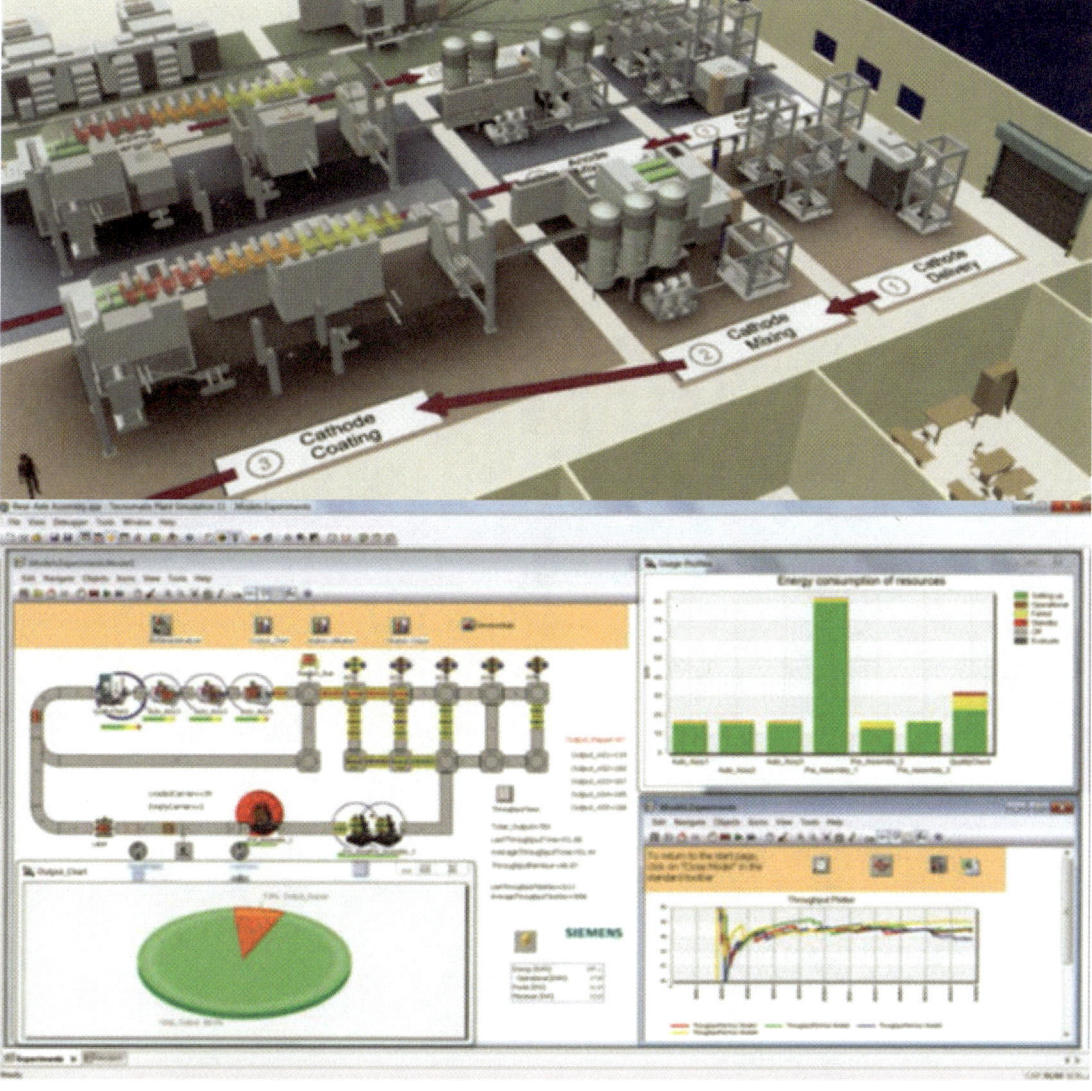

Figure 12.2 Top: 3-D virtual Plant Simulation models that are synchronized at all times with their 2-D counterparts. Bottom: Using Plant Simulation's libraries within a 3-D virtual environment. The virtual 3-D model engine allows flexibility to choose the appropriate method of designing and visualization without compromising simulation and analysis needs. (Images courtesy of SIEMENS AG, 2013.)

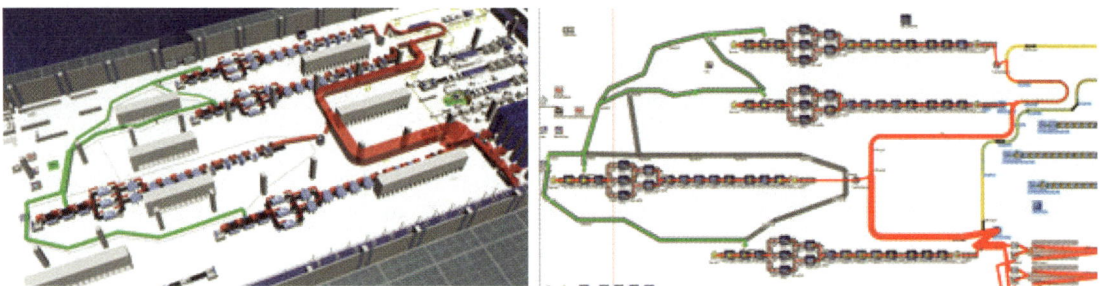

Figure 12.3 Plant Simulation can represent Sankey diagrams in 2-D as well as in 3-D to visualize the flow of material. These Sankey or flow diagrams allows the user to easily read discrete event simulation runs and recognize patterns to draw the right conclusions. The Tecnomatix Plant Simulation Sankey wizard enables users to create Sankey charts just by using drag-and-drop commands in a matter of a few seconds. On the right, images with blue components represent the printed circuit boards (PCBs), and in green the movement of workers through a production plant. In a Sankey diagram, the diameter of the lines shows the quantity or volume of material entity flowing through the system. (Image courtesy of SIEMENS AG, 2013.)

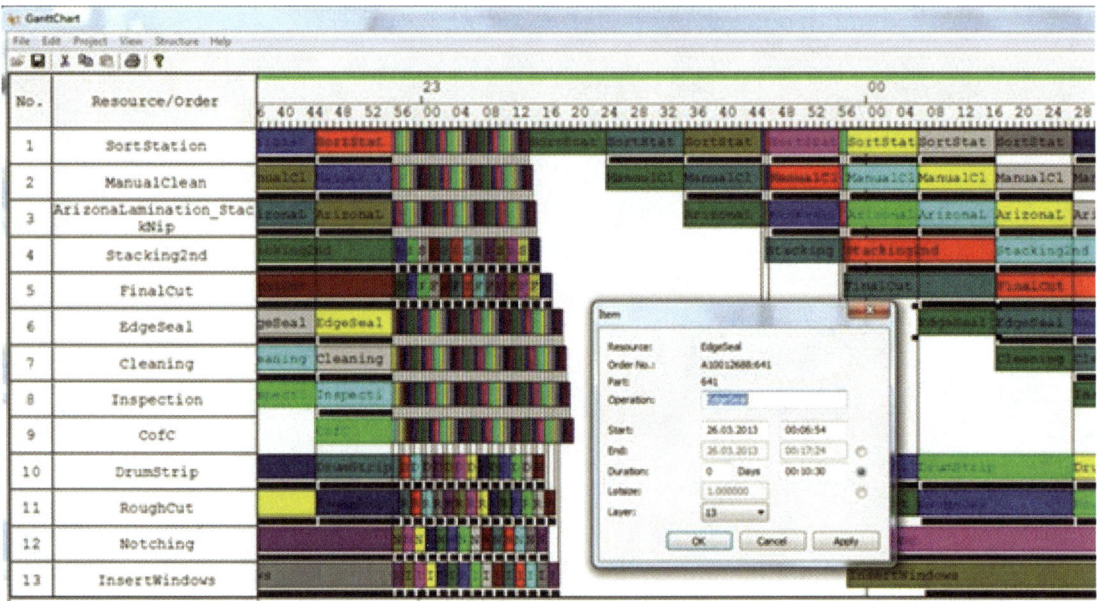

Figure 12.4 Gantt Chart in Tecnomatix Plant Simulation: Plant Simulation Gantt is a program for visualizing and interactively changing sequences of activities. The Gantt chart enables the user to switch between the order view and the resource view plus additional information like failures and shift times. The Gantt chart also provides additional information like beginning time, end time, and product-specific names by double-clicking on one graphical bar. Another feature is the Plant Simulation Gantt wizard, which takes only drag and drop to insert and parameterize the complete Gantt chart. (Image courtesy of SIEMENS AG. 2013.)

let users evaluate different manufacturing scenarios to make fast, reliable decisions in the early stages of production planning. The use of parametric smart objects makes it much easier for teams to understand and represent all of the factory resources (from conveyors, mezzanines, and cranes to containers, to automatic guided vehicles (AGVs) and operators. 3-D virtual modeling helps minimize interpretation errors by enabling team members to see how factory smart objects interact with one another inside a facility [1].

Some of the advanced features are:

- Simulation of complex production systems and control strategies

- Optimize systems for reduced energy usage toward net-zero-energy integrated renewable operation strategies
- Object-oriented, hierarchical models encompassing business, logistics, and production processes
- Dedicated application object libraries for fast and efficient system modeling
- Automatic bottleneck detection and material flow simulation including Sankey diagrams and Gantt charts (Figures 12.3 and 12.4)
- 3-D online visualization and animation (Figure 12.2)
- Integrated neural networks and experiment handling
- Automated system optimization via GAs (Figure 12.3)
- Value stream mapping and simulation
- Open-system architecture supporting multiple interfaces and integration capacities (ActiveX, CAD, Oracle SQL, ODBC, XML, Socket, OPC, etc.)

Case Study: Integrated Tecnomatix and Robotics Process Engineering for the Volkswagen Group

In this case study, we describe how the SIEMENS Teamcenter and Tecnomatix software provides the Volkswagen Group a collaborative environment in achieving digital factory design and built manufacturing flexibility. Both software products are made by Siemens Product Lifecycle Management Software Inc. The virtual factory design, assembly, and vehicle program capabilities help to significantly reduce the development time of the factory. It decreases the number of iteration cycles and prototype phases improving product and process quality and increasing communication of product and process data among departments, worldwide plants, and external engineering partners.

The Volkswagen Group is a multinational automotive company headquartered in Wolfsburg, Germany. It designs, manufactures and distributes passenger and commercial vehicles, motorcycles, engines, and turbo machinery. It also offers related services including financing, leasing, and fleet management. In 2012, it produced the third largest number of motor vehicles in the world, behind General Motors and Toyota [2]. The Volkswagen Group sells passenger cars under the Audi, Bentley, Bugatti, Lamborghini, Porsche, SEAT, Škoda, and Volkswagen brands. It also sells motorcycles under the Ducati brand and commercial vehicles under the MAN, Scania, and Volkswagen Commercial Vehicles marques [3].

For example, Škoda management's digital factory projects included a comprehensive engineering approach to support their industrial production between 2012 and 2013. Škoda management acquired product data, established tool and process libraries from the Volkswagen Group's PLM Software Tecnomatix portfolio, and adopted workflow and data model and assumed offline programming and 3-D robotics process simulation. Škoda's next task was to adopt Volkswagen's Dikab concept (Dikab is a German acronym for digital BIW, meaning body in white). It encompasses a data model, workflow, and methodology that enable improved internal communication within the group and external communication with global

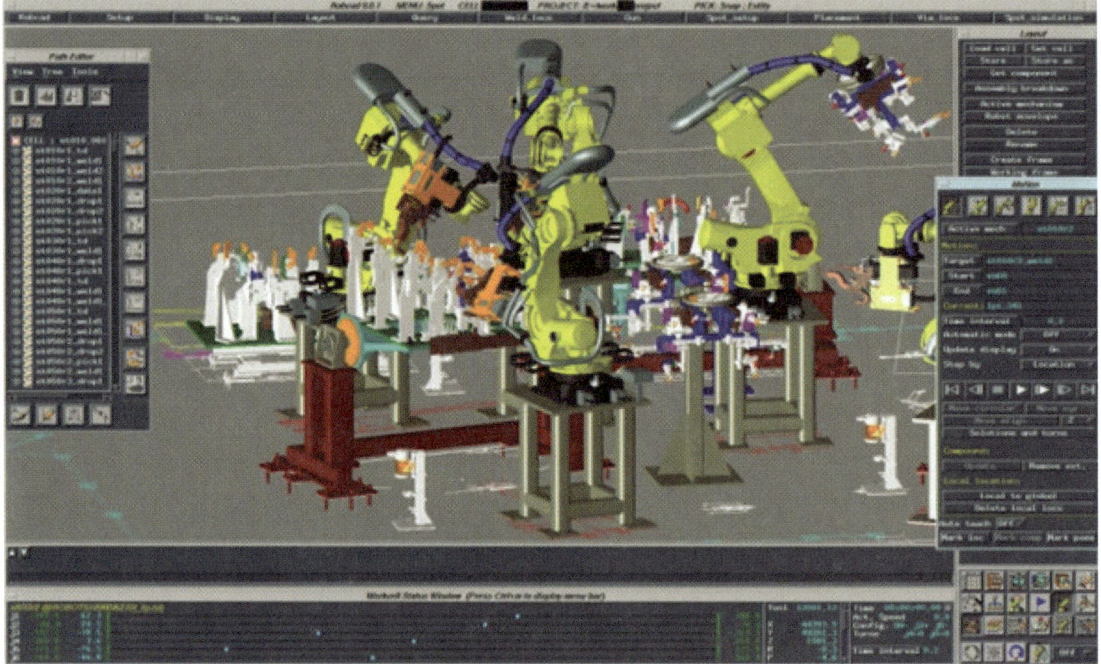

Figure 12.5 Tecnomatix RobCAD: Digital manufacturing environment for robotic work cell verification and off-line programming interoperability with CAD systems Robcad fully integrates with most industry MCAD systems, including native data from Catia, NX™ software, Pro/Engineer, NX I-deas® software, CADDS5, direct CAD interfaces, or neutral formats such as JT™, IGES, DXF, VDAFS, SET, STL, and STEP. (Image courtesy of SIEMENS AG, 2013.)

manufacturing engineering suppliers. It uses Process Designer and Teamcenter software to execute this. Škoda then defined the target of a virtual commissioning project to help shorten the time needed for the production of a BIW line. This is especially important when introducing a new model to an existing, running line.

"Siemens PLM Software's Process Simulate Virtual Commissioning enables us to optimize and troubleshoot both the mechanical and control aspects of a robotics work cell," says Petr Hynek, BIW planning team manager at Škoda. "The value of using Tecnomatix is to connect to a real physical controller, run the simulation, and test different production scenarios by sending signals from the controller is outstanding. It allows for a very thorough validation of the planned robotics work cell interlocks, safety and cycle time optimization. One of the next essential activities was comparison of the digital motion results (recorded using the motion capture suite integrated with Tecnomatix) with the conventional evaluation of the physical motion," says Andrej Bednár from SIEMENS. "We found the digital motion results to be almost the same as the physical motion results, which are impressive. These results reinforced our intention to establish a virtual-reality ergonomics lab in which two engineers from the digital factory team work with Jack and Process Simulate Human, supported by the health protection, product design and manufacturing engineering departments (Figure 12.5) [4]." In the newly established lab, the digital factory team works with state-of-the-art technology. The integrated motion capture Tecnomatix applications use production operation tracking. This allows the teams to analyze the feasibility of work operations, especially with regard to the different heights of workers. Best practices and lessons learned are further used to train shop-floor workers on the basic principle of ergonomics and simulating learning curves with Plant Simulation (Figure

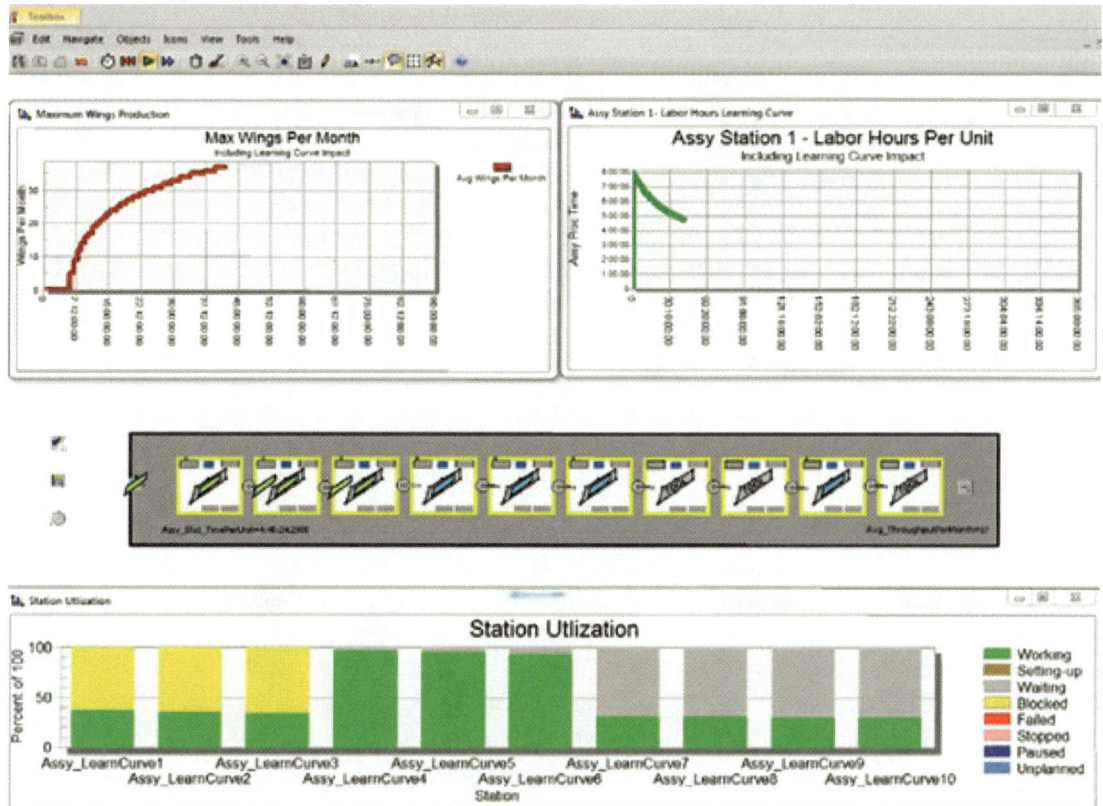

Figure 12.6　Simulating learning curves with Plant Simulation. (Image courtesy of SIEMENS AG, 2013.)

12.6). Expected objectives include improving working conditions for employees, reducing injury occurrences and costs, addressing ergonomics during the preproduction project phases, and digitally verifying assembly operations in compliance with ergonomic standards and laws.

GAs, Neural Networks, and Wasp Swarm Optimization of Logistic Systems and Automation

In this section, we present the optimization of logistic processes in automation and supply chains for digital factory design and operation. Siemens research and development is focusing on neural networks and GAs based on nature. Ants, for example, are fascinating creatures—not so much because they are particularly intelligent on their own, but because in a colony, they display what is known as swarm intelligence. Swarm intelligence can be made use of in logistics and building automation, as Dr. Thomas Runkler from Siemens Corporate Technology explains: "When components of a delivery arrive too late or have been damaged in transit, the warehouse manager has to reschedule orders and decide which one has priority." With today's JIT production, punctual delivery—not too early, not too late—is crucial for companies. Yet conventional logistics programs are inflexible and only reschedule orders, according to a rigid if/then rule. By contrast, Runkler's swarm program operates without any fixed rules. It simply reclassifies the orders and advises the warehouse manager how best to assign an individual component delivery. "It's like ants gathering food," Runkler explains. Initially, they all wander off randomly. But frequently, the shortest route develops more or less spontaneously, as it is where most ants travel and the concentration of pheromones is strongest. This, in turn, attracts even more ants, and a broad "ant avenue" is the result. Runkler's program functions in similar fashion to assign components rapidly and efficiently to individual orders. Meanwhile, when deciding which order should leave the warehouse for a factory production line first, wasps provide a clue. In a colony, each interacting wasp has a specific job—defending the nest, for example, or searching for food to meet the nest's just-in-time demand just like production facilities work. The more important the task, the more resolutely the wasp goes about accomplishing it.

Translated to a mathematical model that employs fuzzy logic, each order corresponds to a wasp. Factors that determine an order's importance include the number of missing components or a possible delay. Once an order reaches the top of the hierarchy, it is dispatched. "In experiments, we have almost perfected the system, with orders being delivered on time in 97 of 100 cases," says Runkler. This improves order punctuality by 50% and practically eliminates the possibility that deliveries can arrive late by 7 days or more. These algorithms have already been used in a number of projects with Fujitsu Siemens Computers and Siemens Industrial Solutions and Services [5].

Another example is the automation of technical processes in manufacturing. In particular, the optimization, procedures, and approaches with artificial intelligence are increasingly replacing traditional methods. Neural networks are frequently applied in such cases, and these have already proven their performance capability in many applications of industrial hardware and software engines. Modeling of technical systems and processes using neural networks allows forecast models to be generated, driving a more target-oriented, increased quality, and uniform process. Significant increases in efficiency are possible using such prediction mechanisms. An advantage is that the simulation of a process requires no detailed knowledge of the prevailing conditions and relationships. Neural networks can also be used for classification tasks and pattern recognition (e.g., in order to identify specific plant statuses). In contrast to classic technical systems, a neural network consists of a complex interconnection of many simple processing units, the so-called neurons. The architecture copies the structure of the biological nervous system. Neural networks are flexible and capable of learning. They can organize themselves and work in a parallel structure.

Typical applicable systems that are capable of learning can be generated by combining neural networks with SIMATIC PCS 7 Add-on FuzzyControl++. This opens up many new possibilities for automation technology in architecture, cities and infrastructures, and product and industrial systems. Neuro systems enable the generation of artificial neural networks in process automation, which cannot be produced using conventional means and methods. Neuro systems can be used to develop neural networks for both simple and effective complex optimization tasks, virtual soft sensors, predictions, identifications, classification tasks, and so forth, even without special know-how. SIMATIC PCS 7 is a system design for process control and programming of total integrated automation. The SIMATIC WinCC SCADA of this product system was successfully used in the Al Bahar Tower in Abu Dhabi as HMI automated control software, which is described in Chapter 5.

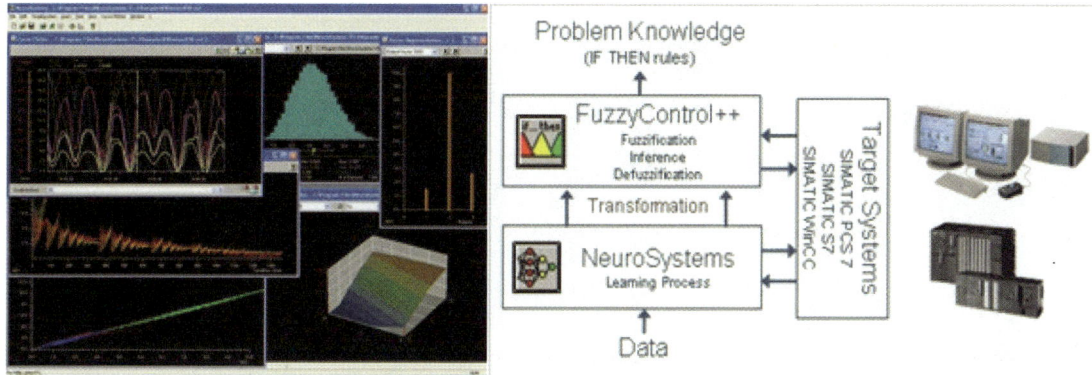

Figure 12.7 Neuro systems are used successfully in the fields of control technology, pattern recognition, prediction, classification and optimization processes. Left: Screenshot of neuro systems. Right: Optimization processes. (Image courtesy of SIEMENS AG 2013.)

In summary, as a result of the complex nonlinear response of neural networks, processes can be simulated without exact knowledge of the existing relationships and conditions. Artificial neural networks have now become the most frequently applied approach for black box modeling of technical, chemical, and biological systems. The capability to learn and adapt, the fault tolerance, and the ability to process inexact or even contradictory information are particularly distinctive.

Neuro systems, as a decision support system for building automation, are well suited to prediction, optimization, classification, identification, and closed-loop control tasks. These applications based on neural networks allow increases in performance, quality, productivity, and efficiency through machine-learning mechanisms (Figure 12.7) [6].

References

[1] SIEMENS AG, Tecnomatix for Plant Simulation, http://www.plm.automation.siemens.com/en_us/products/tecnomatix/plant_design/plant_simulation.shtml, retrieved Oct. 2013.

[2] World Ranking of Manufacturers: Year 2012. World Motor Vehicle Production: OICA Correspondents Survey. Organisation International des Constructeurs d'Automobiles. 2012.

[3] SIEMENS AG, https://eb.automation.siemens.com/mall/en/ca/Catalog/Products/10042041, retrieved Oct. 2013.

[4] SIEMENS AG, Case Study: Skoda creates digital factory, 8 August 2013: http://www.onwindows.com/Articles/Skoda-creates-digital-factory/8084/Default.aspx, retrieved Oct. 2013.

[5] SIEMENS AG, https://eb.automation.siemens.com/mall/en/ca/Catalog/Products/10042041.

[6] SIEMENS AG, NeuroSystems, http://www.industry.siemens.com/services/global/en/it4industry/products/process_control/neurosystems/pages/default_tab.aspx, retrieved Oct. 2013.

Chapter 13

Prefabricating a More Sustainable Building and Assembling It in 15 Days: Broad Group, China

Alfredo Andia

The Origins

In China Broad Group has stunned the category of prefabrication and modular construction by completing several buildings in record time: one five-story construction build in 1 day and a 30-story hotel in 15 days. However, for this company prefabrication is more than just fast construction. It is an endeavor to produce more sustainable buildings and to do so it had to reinvent the whole design-build process.

Broad Group was founded in 1988 by brothers Zhang Yue and Zhang Jian to produce industrial boilers. In the 1990s the company began manufacturing absorption chillers. By 1996 Broad had a 90% of the market share of nonelectric air conditioning systems in China. Two years later, in 1998, the company marketed its products internationally and by 2012 Broad had a "significant market share in many countries in the area

Figure 13.1 T30 Hotel built in 15 days. (Images courtesy of Broad Group, 2013.)

of nonelectric air-conditioning systems, including the United States (an estimated 45% market share), Australia (75%), and India (55%)" [1].

Sustainability Vision

As the nonelectric air conditioning business of Broad Group continued to expand rapidly during the 2000s, the chairman of the company, Zhang Yue, become increasingly more concerned about the environment of the planet. The initial vision of Zhang was that their more energy efficient nonelectric air-conditioning system could replace its electric counterpart in world markets.

With time, Zhang's environmental ambition grew larger. By 2008 China was constructing 500 million square meters of new building per year and using 50% of the world's steel and 70% of the world's cement. Efficient air-conditioning was not really going to tackle global warming at the scale needed.

Zhang become convinced that facing global warming required significantly raising the efficiency of millions of square meters of new buildings that China and the rest of the world were producing. According to Zhang, buildings represent 80% of the world's potential in energy savings. Zhang became Vice Chair of the Sustainable Buildings & Climate Initiative (SBCI) for the United Nations Environment Program (UNEP). In 2009 he was instrumental in drafting the Building Efficiency Guidelines for UNEP, proposing several verifiable indicators for energy efficiency. In 2011 Zhang Yue was a recipient of the 2011 Champions of the Earth prize, the highest award given by the United Nations for environmental leadership.

Broad Sustainable Building

However, the question for Zhang was whether building efficiency guidelines would truly have a chance in impacting the way buildings are made. Zhang came to the realization that Broad Group had to have a bigger role in the actual construction of sustainable buildings if it really wanted to have an impact on global warming. Initially Broad tried to work with architects and engineers but it grew frustrated with the time and cost that process needed. In order to significantly control cost Broad had to challenge the traditional design-construction processes. It had to rethink the entire load-bearing structure and develop buildings that weighed less, used less material, was modular, used less energy, produced its own data, and could quickly assemble on the site. Basically, they had to do an extremely environmentally superior building with significantly less resources, with intelligent performance, and for a significantly cheaper price.

Broad Group had significant experience in factory production with its nonelectrical air-conditioning units. In these factories it began in 2009 a modular prefabrication system named Broad Sustainable Building (BSB) in which more than 90% of the labor-hours of construction are spent in a factory. In 2010 it built a six-story pavilion for the Shanghai Expo in just 1 day and in 2011 assembled and completed the T30 building, a 30-floor hotel, in 15 days. By April 2013 they had already built 25 buildings using this technology, all in China, except one built in Cancun, Mexico.

The T30 Hotel Built in 15 Days

The T30 is a 330-room hotel with 65 parking spaces and is the most famous of the BSB buildings to date. In January 2012 Broad Group released a time-lapse video of the construction of this 30-story tower, which was constructed in 15 days [2]. The video rapidly went viral on the Internet and was widely covered by media around the world. The video shows the 17,338-square-meter and 99.9-meter (328 feet) tower being assembled like a Lego set in just 360 hours (Figure 13.1). The T30 building was constructed inside Broad Group's factory campus in Changsha, China, where the BSB factory already had 220,000 square meters of dedicated workshop floor in 2012.

The T30 was made of several prefab modular components. The basic floor module is a prefabricated steel-framed unit measuring 15.6 m (51.2 feet) by 3.9 m (12.8 feet) with a depth of 0.45 m (1.48 feet). This floor unit weighs 12 tons and incorporates everything from water pipes and electrical systems to ventilation shafts, ceiling lights, sinks, floor finishes, and toilets. Other components produced in the factory are the vertically braced columns, interior walls, and all the façade modules. The T30 building has a span distance between columns of 7.8 m (25.5 feet) and a clear height of 2.75 m (9 feet).

Figure 13.2 Truck with two floor modules. All the vertical elements that need to be assembled on the job site are laid out flat on top of the floor module ready to be raised with minimum effort once the panel is lifted into its position. (Image courtesy of Broad Group, 2013.)

Figure 13.3 The floor modular panels with all the vertical elements laid on top are speedily removed from the truck and raised to their location and bolted to the braced columns. No welding is used on the site (Image courtesy of Broad Group, 2013.)

All these elements are designed to be moved via trucks into the job site (Figure 13.2). One truck can carry 120 square meters of prefabricated floor modules. The whole site material delivery was carried in 150 vehicles. Broad states that with a comparable conventional building would have taken around 1,000 vehicles. At the assembly site the independent floor modules units are raised from the trucks into their location with all their vertical components laid flat in the unit. The floor modules are positioned next to each other guided by their matching connecting plates. The electrical and mechanical systems are quickly linked between the floor modules and the vertical elements such as columns and interior partitions are all raised and bolted into place. No welding is used on the construction site (Figure 13.3). Once several floors of the modular floor units and vertical elements are bolted in place the prefabricated façade system components are placed to enclose the building.

BSB Sustainability

Broad Group in the T30 Hotel define sustainability in eigth aspects [3]:

1. Earthquake resistance: A building that resists a magnitude 9 earthquake.
2. More energy efficiency: Broad claims that its building is approximately five times more efficient than a traditional building by using 30 different energy saving strategies and technologies. These strategies include very high thermal insulation in the exterior walls and roof, heat recovery fresh air, multipaned windows, solar shading, LED lights, water-saving toilets, and energy recovery elevators.
3. Air purification: Twenty times more purer air by creating low-cost super filtration equipment and by installing air quality detectors in each room for indoor PM (PM0.3, PM2.5, PM10), formaldehyde, and CO2 levels that are compared with the exterior PM.
4. Durability: Broad claims that the building is designed to last 600 years (with inspection and maintenance every 60 years).
5. Material savings.
6. Recyclable construction.
7. Recyclable materials.
8. Clean materials: Construction materials with no formaldehyde, lead, radiation.

Broad asserts that the total energy consumption per year (per primary energy) of the T30 hotel is 2.2 million kWh. It states that this represents 20% of the energy consumption of a comparable traditional building (including five-star hotels), which according to its calculations will consume 11 million kWh (Figure 13.4).

Cost and Time

According to Juliet Jiang, senior vice president of Broad Group, the prefabrication of 93% of the T30 took 45 days in Broad Group's factory [4]. She said that now it could finish the prefabrication production of a

No.	Category	Items	ⓑ BSB	Traditional buildings (including 5-star hotel)
1	Key index	A/C and ventilation energy consumption (per primary energy)	70kWh / m²a (equivalent to 7 kg oil)	350kWh / m²a (equivalent to 35 kg oil)
2		Average heat-transfer coefficient of building envelop	0.3W / m²K	2W / m²K
3		Power distribution for lighting (on average)	2W / m²	6W / m²
4		Toilet water consumption (each time)	3 liters	12 liters
5	Thermal insulation	Materials of external wall thermal insulation	Rock wool 150mm 0.23W/m²K (inside glass curtain wall)	Little or no thermal insulation
6		Window, glass layers	4 layers	1 or 2 layers
7		External solar shading	Automatic shutters (in glass curtain wall)	Internal solar shading
8		Internal window thermal insulation	Automatic curtain	No
9	Ventilation	Ventilation equipment	Heat recovery fresh air machine	No heat recovery
10		Ventilation power consumption	0.6~0.9W / m³	1.2~1.8W / m³
11		Fresh air heat recovery efficiency	70~90%	No
12		Fresh air by pass	Air does not go through heat exchanger during transitional seasons	No
13		Air supply method	Underfloor air supply	Ceiling air supply
14		Fresh air flow route	7~15m	3~5m
15	Equipment	Chiller/heater	Non-electric air conditioning Total COP 112%	Electric air conditioning Total COP 52%
16		Power consumption of A/C water distribution system (electricity/cooling)	3%	10%
17		Room temp. regulating methods	Central fan coils (2 sets for entire building), Mix of fresh air & exhaust air can be adjusted automatically in each room	One set of fan coils for each room
18		Indoor humidity regulating methods	High pressure water mist	Steam
19		Elevator	Generate power when ascend empty-loaded or descend fully loaded	No power generation
20		Kitchen ventilation	Inverter controlled	Fixed
21		Laundry drier	Waste heat from Chiller/heater & power generation	Steam or electricity
22		Drinking water	Produced by hotel itself (reverse osmosis water)	Outsourced bottled water
23	Smart control	Fresh air & air conditioning	Automatically turned off 2 hours after people's departure	No
24		Fan frequency regulating	Inverter controlled	Fixed
25		External solar shading	Automatically start when temp. ≥23°C	No
26		Internal thermal insulation curtain	Automatically closed when temp. ≥33°C or ≤14°C (no people inside)	No
27		Lighting in rooms	Automatically turned off half an hour after people's departure	No
28		Lighting in public areas	Automatically turned off when people leave	No
29		Energy metering	Independent metering, total metering	Total metering
30	Others	Lighting source	All LED lighting (100 lumen / W)	Incandescent or fluorescent lighting (10~70 lumen/W)
31		Garbage classification, recycle	8 garbage shafts for each floor	No
32		Recover heat from bathing waste water	Heat up tap water in winter	No
33		Utilize toilet sewage water	Produce biogas	No
34		Thickness of A/C water & hot water pipe thermal insulation	80mm	20 mm
35	Total energy consumption a year (per primary energy)		2.2 million kWh	11 million kWh

Notes: 1. Energy consumption, thermal insulation & window layers are per the standard of "Hot summer & warm winter areas". See the comparison list in the following page for other climate areas.
2. Calculation basis: Converted primary energy/electricity: 4kWh/kWh, converted oil/electricity: 0.25L/kWh, annual lighting hours: 2000, hotel occupancy rate: 80%.
3. Compared with traditional buildings (including 5-star hotels), this hotel saves 8.8 million kWh a year in terms of the total energy consumption of air conditioning, ventilation, lighting, elevators, water pumps, etc. If we convert it into oil per primary energy (10kWh/L), it equals 880,000 liters or 730 tons of oil saving and 2000 tons of CO_2 cutting each year, which is equivalent to 110,000 tree planting.

Figure 13.4 Energy conservation comparison list [3]. (Image courtesy of Broad Group, 2013.)

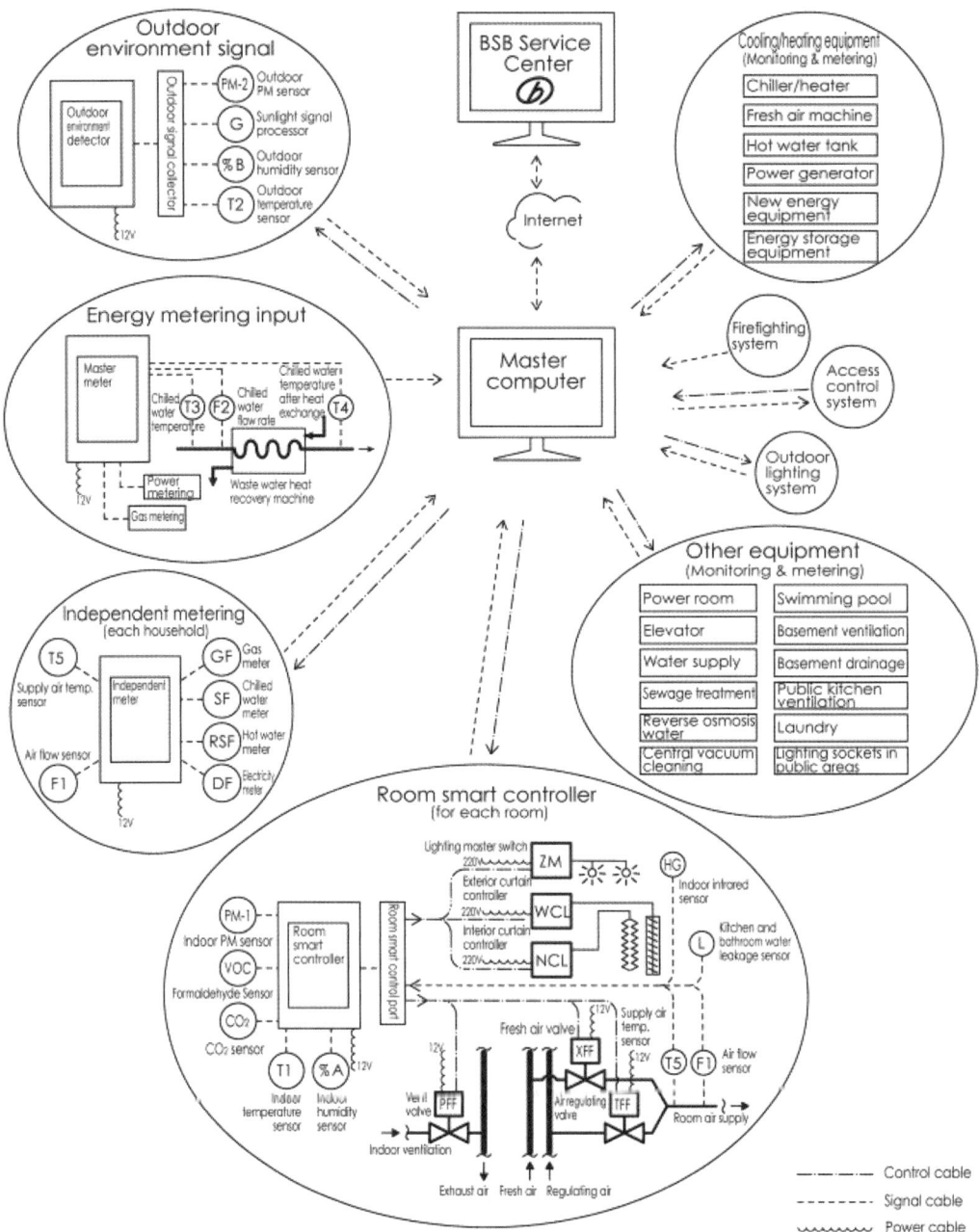

Figure 13.5 Diagram of the Building Automation System (BAS) for the T30 Building [3].

building similar to the T30 in just 7 days. Ms. Jiang said that the cost of the T30 Hotel was US$1,000 per square meter. She adds that the price is inflated because it was the first prototype. Ms. Jiang estimates that a second building similar to the T30 will cost around US$500 per square meter.

The construction time of 15 days of the T30 does not include the foundations. The foundations of the T30 are conventional concrete foundations. The only difference in the foundations is that they are smaller since the building weighs less. Once the modules arrived at the site the T30 hotel was assembled in 360 hours or 15 days. The total time from ground breaking to opening of the 700-bed hotel was 48 days. The construction conditions were also extraordinary—on site there was no welding, no water, no injuries, and only 1% waste compared to traditional construction.

Building Automation System

The BSB has all its monitoring systems connected to a master computer and to the BSB service center via the Internet. This service center connects to all the building systems such as cooling/heating equipment, firefighting, access control, outdoor lighting, power room, elevator, water supply, sewage treatment, and basement drainage. It also connects to the sensors, controllers, and metering system in every room and to the outdoor environmental signal that monitors data such as PM2, sunlight, humidity, and temperature.

Business Model

According to Broad Group senior vice president Juliet Chang, the business model is to license the technology worldwide to local franchises. Broad Group's plan is to create 150 franchises, 50 in China and 100 abroad. The plan is to do 93% to 95% of the work in a local factory and only 5% to 7% of the labor on site. By 2012 Broad had six BSB franchise partners in Ningxia, Fujian, Shangdong, Shanxi, Henan, and Hubei, China.

Sky City: The Tallest Building in the World Built in 90 Days

BSB's most ambitious project to date is to construct a 220-story building called Sky City in Changsha, China. Sky City will be the tallest building in the world, house 30,000 people, and have 17 helipads. The building is intended to mix all types of programs where people can live, work, educate, shop, and entertain. They are projecting that they will need a workforce of 16,000 people in order to prefabricate the modules in the factory for 6 months. The whole project is intended to be assembled on site in 90 days.

Conclusion

Just constructing a 30-story building in 15 days is a major achievement in the history of construction, and doing it 25 times in 3 years begins to build a track record. BSB's main workshop is being planned to increase to 360,000 square meters and house 19,000 workers to produce 10 million square meters of buildings per year. Beyond these records the BSB building is a lesson on how the construction industry is beginning to profoundly transform the design and construction process to meet the enormous global challenges we are facing. It is an exercise in radically transforming the way we think about the discipline of design and building by redefining its most basic principles.

References

[1] E Jacqui Chan. "City&Country: Broad Group's sustainable buildings making waves." The Edge Malaysia, Sun, Jul 8, 2012.

[2] Video. Broad Group. History Rewritten (T30 Hotel) (3'03") http://en.broad.com/video.html?3.

[3] BSB. Broad Sustainable Building. T30A Tower Hotel. January 2012. http://www.greenindustryplatform.org/wp-content/uploads/2013/07/Broad-Group-BSB-T30-Tower-Hotel_Technical-Briefing.pdf.

[4] Building 30 Stories in 15 Days: Modular Construction with the BROAD Group. Interview by Cesar Abeid, April 4, 2013. http://www.remontech.com/building-30-stories-in-15-days-modular-construction-with-the-broad-group/.

Chapter 14

Automated Fabrication and Assembly: Sekisui Heim, Tokyo, Japan

Jun Furuse, Masayuki Katano, and Thomas Spiegelhalter

Introduction

In Japan, most prefabricated houses are manufactured by using automation, service robotics, and other advanced assistance technologies. The steady success of the main players in Japan's prefabrication industry like Sekisui Heim (house), Daiwa House, and Toyota Production System (TPS) demonstrate decades of reliable products and services in the mass customized housing kit industry [1].

In the 1980s, Sekisui Heim developed through their parent company Sekisui Chemical Co. Ltd. worldwide an innovative, "computer-based Enterprise-Resource Planning (ERP) system for controlling the production and logistics flow"[2]. In the housing manufacturing section this ERP system from the 1980s was developed further toward the Heim Automated Parts Pick-Up System (HAPPS) [3]. The system translates design parameters and plans from architects, engineers, and clients directly into parametric-algorithmically processed production plans and data needed to operate fully automated production.

Sekisui has one of the world's most advanced 3-D-/4-D BIM allowing more than 90% of all generic or individual designs and parts-related information to be directly translated into production and assembly operations. Sekisui incorporates in its total building factory automation as well dynamic data management abilities for the life-cycle of its building products, long-term customer relations and after-sales support, durability and maintenance performance, barrier-free-, earthquake-, wind-, and fire-retardant resistance, zero-emission upgrade offers, solar energy generation systems, and rearrangement, deconstruction, and reuse or recycling of the low-energy Sekisui homes [4]. Sekisui Support System provides housing customers with a 60-year scheduled diagnosis support system for Heim Products and a 20-year warranty system for Heim's superior building frames, walls, and airtightness and water tightness. The warranty conditions involve carrying out scheduled 10- and 15-year inspections [5].

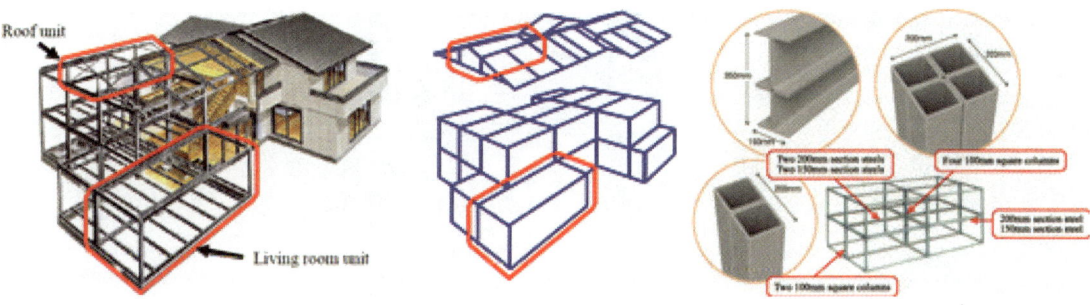

Figure 14.1 Unit method with steel structure frames. (Source: Jun Furuse and Masayuki Katano, Sekisui Heim, 2006.)

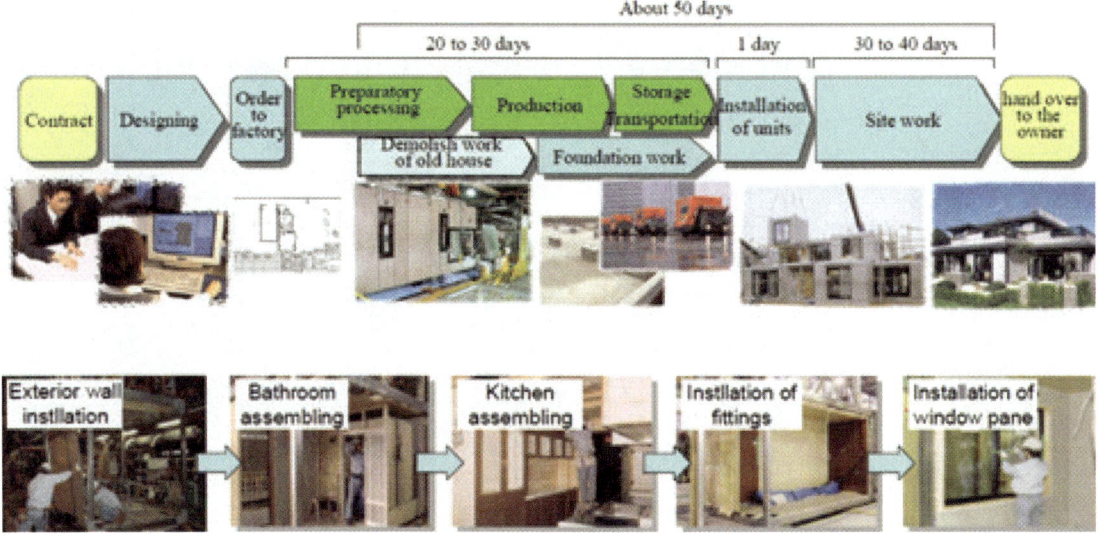

Figure 14.2 Workflow of the Heim manufacturing and factory assembling of units. (Source: Jun Furuse and Masayuki Katano, Sekisui Heim, 2006.)

The following case study will describe how various house designs are parametrically translated and algorithmically configured into unique automated pick-up parts consisting of prefabricated Heim units. It will describe and summarize the automated just-in-time methods and processes, and the efficiency and accuracy of HAPPS.

Modular Sekisui Unit House

The high-quality modular Sekisui Heim residential houses are made of factory-produced modules called units. These steel-frame units are manufactured using the unit construction method, which employs factory production to build homes precisely according to common or individual design specifications. Sekisui claims that the units achieve the highest levels of performance and life-cycle standards according to Japanese building benchmarking. All units of a house can be uniquely fabricated

Figure 14.3 Automated workflow of the Heim manufacturing and factory assembling of units with robots. The robots position automatically various units depending on the steel component configuration. Every 3 to 5 minutes a new basic frame element is generated for the next step of wall production. (Still images: Sekisui Heim, Building my house: https://www.youtube.com/watch?v=KWPlNA6DhiM, accessed on Oct. 6, 2013.)

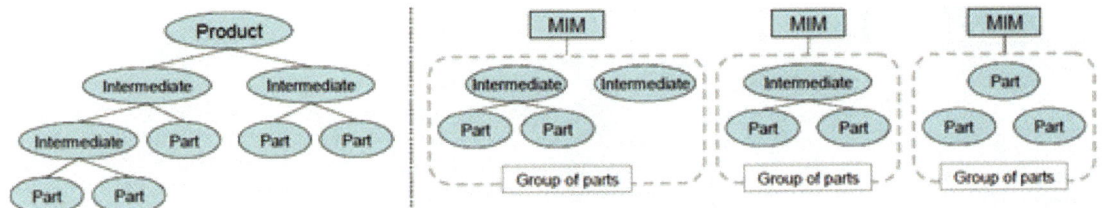

Figure 14.4 Left: Bill of parts of the assembling process. Right: BOM of Heim. (Source: Jun Furuse and Masayuki Katano, Sekisui Heim, 2006.)

with different parts in different combinations. The automated design-to-factory file process can involve the selection and pick-up of about 30,000 parts for each house. The available parts system contains about 300,000 listed parts and can be fed just-in-time to the automated and robot-assisted production line.

Sekisui builds and sells approximately 13,000 Heim unit houses annually including the Zero-Utility Cost House with solar energy generation systems. The units have steel frame structures and are prefabricated in the factory to 80% completion. This process is called unit method and minimizes the work-load at the construction site and enables fast and high-quality house-making. One advantage of the unit method is the autonomy in floor plan development and selection, which is possible through the strong and easy-to-configure steel frame structural systems. Floor plans of contracted Heim houses are divided into several standardized segments for the production for best-suited units (Figure 14.1). There are about 70 kinds of units, of which 40 are standard cuboids varying in 10 lengths, 2 widths, and 2 heights, and others are special shape units (trapezoid, etc).

Workflow from Client's Design Contract to Manufacturing with Robots to On-Site Assembly

Once the customer places an order, each house is processed separately. First, architects or the sales company draws the floor plan, and then at the factory all necessary parts are picked according to the floor plan. The unit frames are made from steel stocks by welding and fed to the assembly conveyor line with automated robots where the installation of exterior and interior parts and equipment to the frames is performed by workers (Figures 14.2 and 14.3). Lastly, the units are transported to the construction site and assembled to compose a house. The large Sekisui factory is able to produce about 135 units a day (1 unit every 3 minutes),

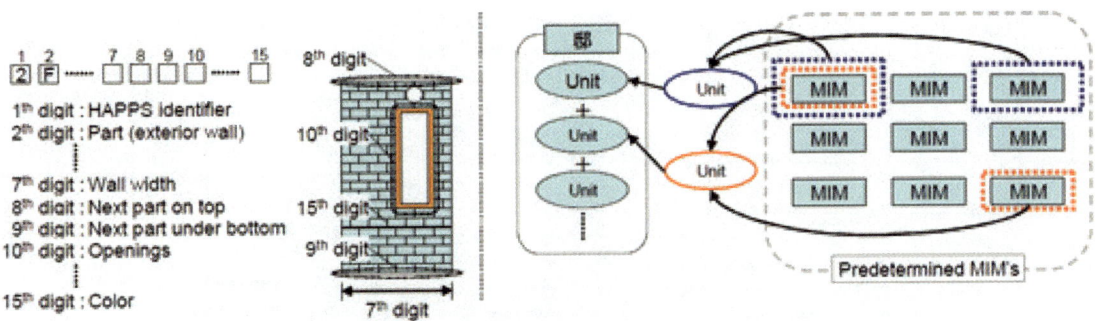

Figure 14.5 Left: Identification of intermediate by menu item master (MIM) code. Right: Image of a MIM pickup. (Source: Jun Furuse and Masayuki Katano, Sekisui Heim, 2006.)

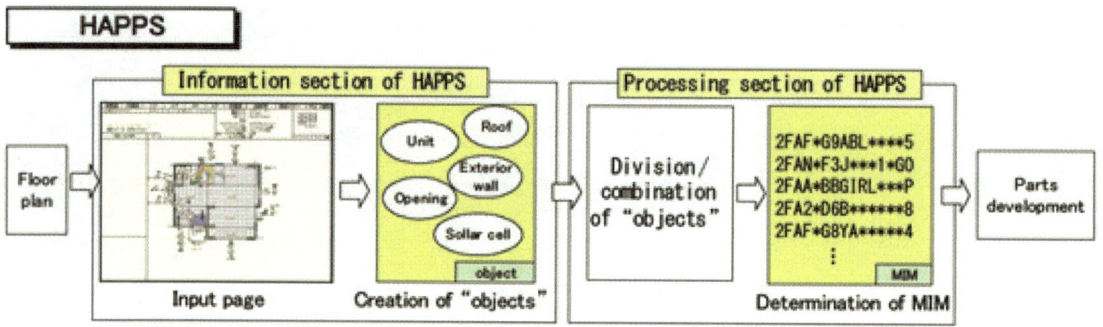

Figure 14.6 HAPPS. (Source: Jun Furuse and Masayuki Katano, Sekisui Heim, 2006.)

each of which consist of different parts placed in different locations. It is important to select about 30,000 specific parts from about 300,000 available listed parts and deliver them just-in-time to the assembly line. The process is similar to the robotics-driven automotive industry: Engine, seat, paint color, audio component, and so forth may be chosen in the assembly of automobiles, but even the layout of seats or windows or size and shape of body can be individually selected in the assembly of Heim. The workflow in Figure 14.2 shows the entire process from contracting and designing to the final assembly.

Technical Key Points in the Arrangement and Programming of Parts

There are two technical key points in the system. One is how to incorporate the bill of materials (BOM) for ultrahigh-mix low-volume production to give design freedom to customers who demand individual houses with different floor plans, equipment, and interior decoration. The other point is the individual arrangement and programming method that breaks floor plans down to parts and groups of parts (Figures 14.4 and 14.5).

BOM Structuring

In the assembling industry, BOM is created by arranging the product into intermediates and then into parts (Figure 14.4, left). The concept of BOM is based on the standard model to which optional parts are tied. There are no standard houses in the Heim offered, because individual floor plans would vary too much. To deal with this situation, a concept of imaginary intermediate is created, which is a group of closely related parts

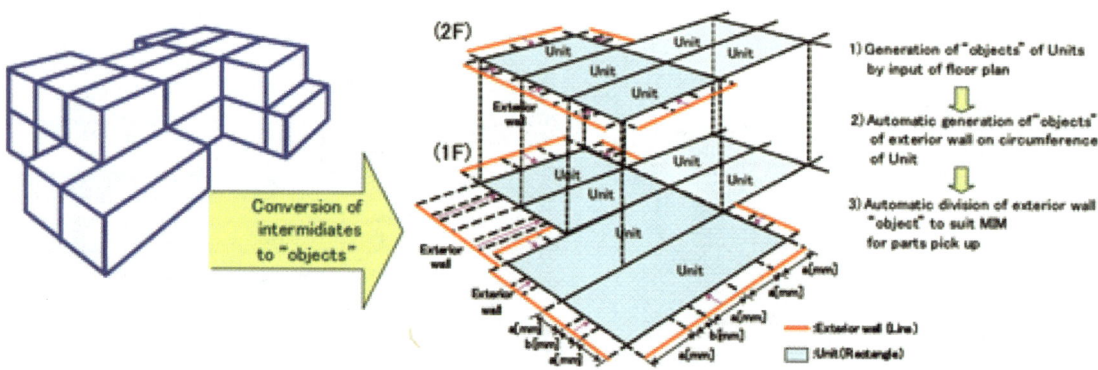

Figure 14.7 Conversion of intermediates to objects. (Source: Jun Furuse and Masayuki Katano, Sekisui Heim, 2006.) Masayuki Katano, Sekisui Heim, 2006.)

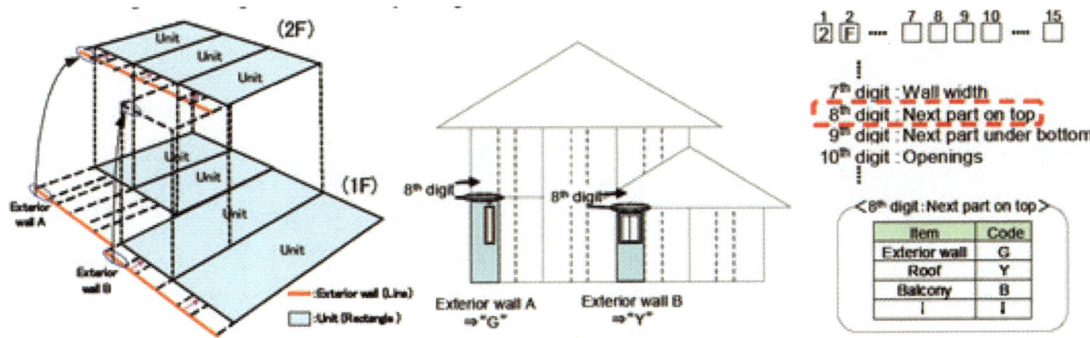

Figure 14.8 Objects and example of MIM determination. (Source: Masayuki Katano, Sekisui Heim, 2006.)

(Figure 14.4, right). BOM is tied to a group of parts and plays an important role in processing individual floor plan configurations. These imaginary intermediates may exist temporarily in the production process or can just be dummy placeholders. They are defined as intermediates, whether real or imaginary, and can be reused for optimized product programming and arranging of parts and systems. Identifying codes of these groups of parts are called menu item master codes (MIM) (Figure 14.5, left). MIM code consists of many digits that indicate, for example, an exterior wall, dimensions, neighboring parts, interfacing condition, color, and so forth (Figure 14.5, right).

Part Arrangement System and Outline of HAPPS

BOM and MIM technologies are organized based on the actual level of research and development (R&D), and there are about half a million MIM codes per one series of the Heim unit products. Orders from customers are individually processed by picking and combining about 4,000 MIM codes from half a million codes on average per house (Figure 14.5, right). The selected MIMs are matched to BOMs Heim production methods which

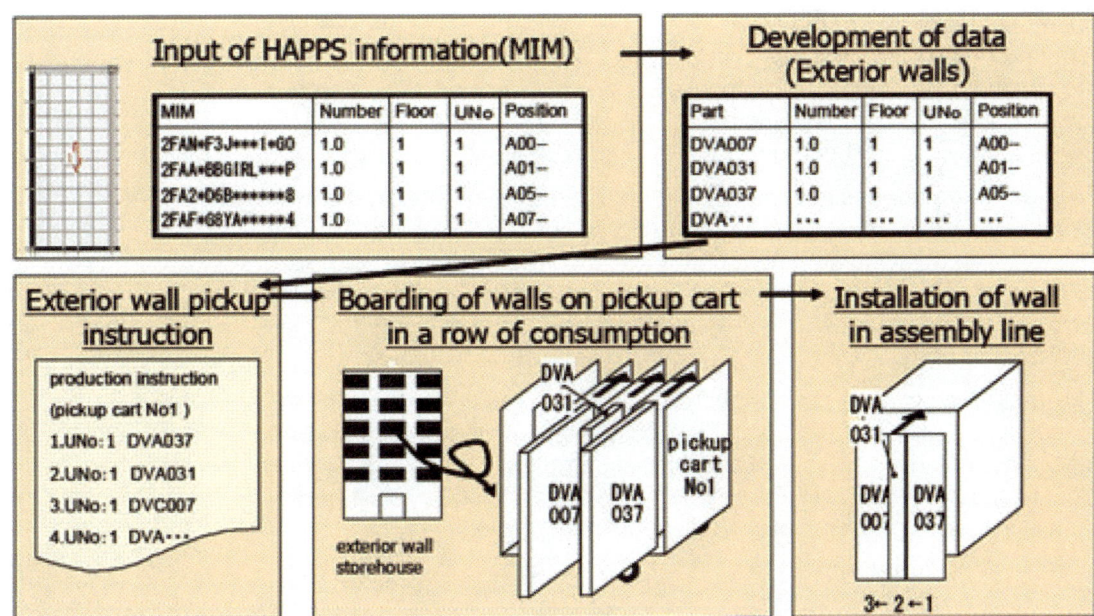

Figure 14.9 Application of HAPPS information in exterior wall installation.

are then designed for the final arrangement and configuration of the parts. This efficient and accurate system of MIM pickup is called HAPPS.

The outline and the important technical points are explained below: HAPPS is a system to pick up and process necessary parts when floor and section plans of the production order with specific information are placed (Figure 14.6). In the information section, input of floor plan data in graphic form, such as house type (snow accumulation class, exterior wall, color, etc), Units, accessories (balcony, entrance porch), and equipment (bath, kitchen, technology room, etc), are converted to objects. The object is abstracted floor plan information expressed in dots, points, lines and rectangles. Objects compose a multidimensional modeled house in the virtual space (Figure 14.7).

In the processing section, objects are restructured (divided or combined) that to new objects suit the MIM's codes of the pickup system. Objects are principally in one-to-one correspondence configured to MIMs codes (groups of parts). Objects must meet and connect to corresponding MIMs and determine the number or letter in each digit in consideration of the surrounding conditions. Figure 14.7 shows a house expressed by units and exterior walls and its conversion to a modeled house composed of objects in the virtual space.

In HAPPS, it is an important process to convert intermediates to objects in the virtual space and to let the objects conform to each other, prior to the pickup of the correct MIM codes.

Conversion of Intermediates to Objects

Based on the location information within the unit and interference information with other parts, objects are allocated to compose an abstract house. The location of the actual intermediates is required for the process of the parts pickup. Corresponding to the allocated object, the MIM codes will be the output referring to the surrounding conditions and selected figure or letter of each digit. The conversion of intermediates to objects optimizes the output of the MIMs very efficiently, which otherwise would require complicated processes of hierarchical selection. In Figure 14.8, the configuration of the exterior top wall (the eighth digit of MIM) is shown as an example: There is an exterior wall on top of the exterior wall A. Consequently, the next part on top is an exterior wall.

There is no exterior wall nor unit but there is a roof on top of the exterior wall B. Therefore the next part is roof. This information and other surrounding circumstances including interfering information of the next part are required to identify the specific MIM codes. Similarly, other digits such as next part under bottom or openings are determined in relation to the units and to other parts.

The mutual composed positions of objects in the modeled house in the virtual space become clear, and MIMs can be identified accurately and efficiently. In addition, objects make the programming easier by generating clear images of the part's fit. Furthermore, the visualization by using objects makes the part's programming significantly more efficient.

Property Inheritance from Object to Parts

About 10 new Heim models and about 400 modifications are launched annually. It is necessary to modify or add to the program in response to such changes by exactly defining the alternating work process. This means the program must always be flexible and durable for alteration or addition.

The HAPPS program is made durable for alteration by providing inherent relations between objects and subordinate parts. This type of relation enables a reduced effort in writing or scripting additional changes within the superordinated program. In this way, it is easy to find related points, which minimizes the error rate of the programming.

For example, if a new color is added to the house, it is scripted as an attribute the new whole house. The attribute information is then given to the selected exterior wall, openings, balcony, and so forth. However, this attribute of wall color is not given to objects of these intermediates, but inherited within all the other house attributes. In case of an addition or change of specific wall colors, the modification of the program is necessary only within the house information of attributes. The inheritance plays a role in reducing the writing or scripting volume and error rate of modification (Figure 14.9).

Application Scope of HAPPS Information

Each object of HAPPS has location information of its own intermediate and interfacing information to neighboring intermediates. This information is also used for parts ordering and production instructions. Integration of information from HAPPS and information of the production scheduling will generate instructions for workers and machines in binary or printed form. Such a function of HAPPS enables just-in-time, in-right-quantity feed of parts for assembly lines. As parts can be exactly scheduled for the expected demand, so a flexible and highly adaptive working environment can be provided for assembly-line workers, who then just pick up the parts at the end of each lineup.

Summary of the Efficiency and Accuracy of HAPPS

The above-mentioned HAPPS is not the only systemic method Sekisui is using, they use other systems for parts programming as well. HAPPS covers about 70% of all the important parts of a house, including structural frames. HAPPS parts programming takes about 1 to 1.5 hours per house depending on the size and specifications, starting from the input of floor plan and ending in the inspection review. The total hours of parts programming of a house is between 5 to 6 hours.

The HAPPS program receives regular feedback from the production team and is therefore updated and released once a month. Before the release happens, the systems check requires an input of about 70,000 test floor plans. For example, if there is a model change, the current MIMs are checked for malfunctioning that may be caused by the program change. If a new model is added, test floor plans are also added to make sure that the program generates correct solutions. In addition, if a new series of Heim is launched, the parametric-algorithmic simulation check is conducted before the release. The benchmarked rate of the correct responds by the MIM pickup is about 99.5% in all conducted simulations. In 2005, the error rate in the actual operation was 0.017 errors per house. This was calculated from 223 errors in about 13,000 Heim houses.

In summary, HAPPS has been very helpful to increase the overall accuracy of the part pickup process and to decrease picking time at Sekisui, thus contributing to the efficiency of part handling. In the future Sekisui will further minimize the error rate by improving the parametric-algorithmic definition and of MIM and the parts grouping method for optimized production.

References

[1] Poorang A.E. Piroozfar and Frank T. Piller. *Mass Customization and Personalization in Architecture and Construction.* (Routledge, 2013).

[2] Sekisui Housing Company. http://www.sekisuichemical.com/about/division/housing/index.html, accessed Oct. 6, 2013.

[3] Jun Furuse, Masayuki Katano. *Structuring of Sekisui Heim Automated Parts Pick Up System (HAPPS) to Process Individual Floor Plans.* (ISARC), (2006), pp. 352—356.

[4] Sekisui Housing Company. http://www.sekisuichemical.com/about/division/housing/unit/index.html, accessed Oct. 6, 2013.

[5] Sekisui Housing Company. http://www.sekisuichemical.com/about/division/housing/support/index.html, accessed Oct. 6, 2013.

Chapter 15

Customized Prefabrication in Two Hospitals: NBBJ, Ohio

Alfredo Andia

Introduction

The ideas of prefabrication and modularization are becoming integrated into processes of design and construction as architects, engineers, and contractors embrace more sophisticated BIM technology and more integrated project deliveries. BIM has allowed professionals to visualize in advance assemblies, conflicts, construction schedule, and cost estimation. Today an emerging number of designers, contractors, and subconstructors that use a BIM model have moved into prefabricating many of their building components.

The Columbus, Ohio, office of architectural firm NBBJ has developed two hospitals that are an important international example on how BIM and prefabrication are intersecting. Hospitals are among the most difficult type of design typology. They have exceptionally intricate programs, use complex equipment, and deal with very delicate types of human factors. So for Tim Fishking, a principal at NBBJ, when they began to think about prefabrication they were clear that they wanted their projects to be driven from an aesthetic and functional standpoint and not defined by a purely prefab strategy. This meant customizing prefabrication solutions for NBBJ's unique health care clients.

Miami Valley Hospital: Implementing the Idea of Prefabrication

The first project in which prefab was implemented by NBBJ was in The Miami Valley Hospital Heart and Orthopedic Center in Dayton, Ohio. The building is a 480,000 square-foot, 12-story addition, $137 million facility for regional diagnostic and treatment services. They saw the possibility of integrating prefab ideas into this project when Tim Fishking, AIA, NBBJ's principal-in-charge, and Marty Corrado, project executive for Skanska USA, met during one of the first planning meetings. Skanska had a serious interest in developing this type of project in the United States as the construction firm was working at that time on two hospitals in the United Kingdom that were using prefab intensively.

Once the design was defined, NBBJ and the hospital construction manager Skanska Shook, a joint venture of Skanska USA and Shook Construction, began to look at what systems were most suited for modularization. They defined five prefabrication and modularization initiatives: (1) Unitized exterior curtain wall, (2) modular and demountable caregiver workstations through the inpatient corridor, (3) the inpatient room with its case work, headwalls, and patient bathroom, (4) the integrated MEP racks above corridors, and (5) temporary pedestrian footbridge.

1. *Unitizing the exterior curtain wall.* A prefabrication and modularization strategy for curtain walls is not a new concept, but for this project it was a critical factor given that the job site did not have a large lay-down area. The

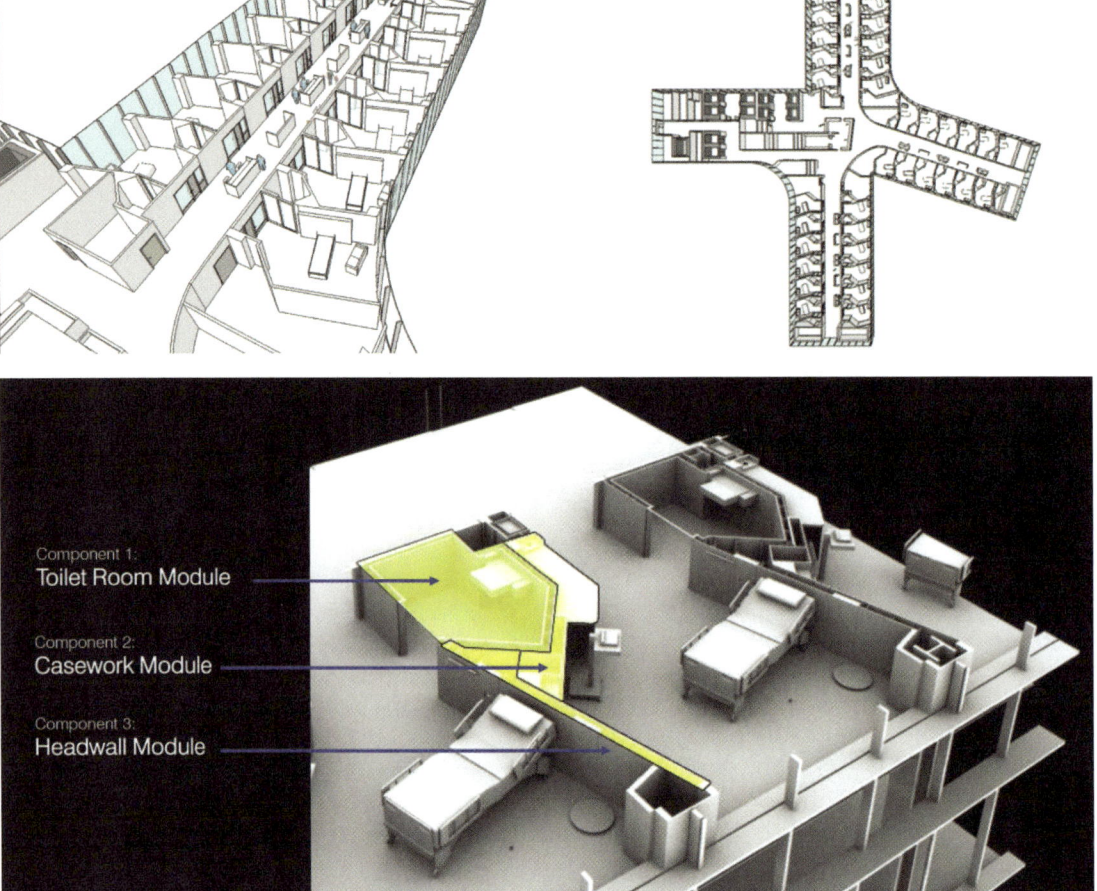

Figure 15.1 The images at the top depict the 12 rooms per wing. The image at the bottom shows the two different prefab components that created the inpatient room. (Image courtesy of NBBJ.)

exterior wall system was designed to be modular but with a fairly random pattern. To increase closer coordination the architects developed a design-assist relationship with the fabricator to involve the fabricator early on in the design process.

2. *Modular and demountable caregiver workstations.* The inpatient hospital tower designed by NBBJ had seven repeated floor plans in the hospital tower and three inpatient wings per floor. Traditionally this is repetitive build-in case work that becomes very inflexible if it has to be modified in the future. In this case a corporate system furniture solution was used to allow the client to make changes inexpensively.

3. *The modularization of the impatient room with its casework, head walls, and patient bathroom.* A typical floor has three wings with 12 beds each as depicted at the top of Figure 15.1. The first instinct was to prefabricate the 180 rooms; however, each room was 22 feet deep and 15 feet wide, which created a serious problem for transportation. So instead it was decided to prefab walls that separated the rooms into components, as shown at the bottom of Figure 15.1.

The architects looked for prefab modular bathrooms in the market but they were not able find any that could fit their standards. Skanska Shook

Figure 15.2 The bathroom pod being prefabricated and loaded into a standard flatbed truck. (Image courtesy of NBBJ.)

Figure 15.3 Bathroom pod being lifted and dollied into its location, which is marked with a 3/4-inch depressed notch in the concrete. (Image courtesy of NBBJ.)

rented a warehouse near the job site to build these modules but before they advanced in the production they built full-scale mock-ups that were tested and critiqued by the hospital medical staff, which gave significant input with regard to the location of fixtures and maintenance issues. Full-scale mock-ups of the inpatient rooms and MEP racks were also critical for the bidding process and for regulatory agencies to understand the process.

Precision and coordination in the delivery process was critical. The construction crew built dollies to position the three prefabricated components when they arrived at each floor. A slightly sunken notch in the concrete of the patient floors indicated exactly the location where the bathrooms should be located (Figure 15.3). It took around one 8-hour workday to install 33 bathroom pods and only 1.5 weeks to rough-in a 30,000-square-foot patient floor.

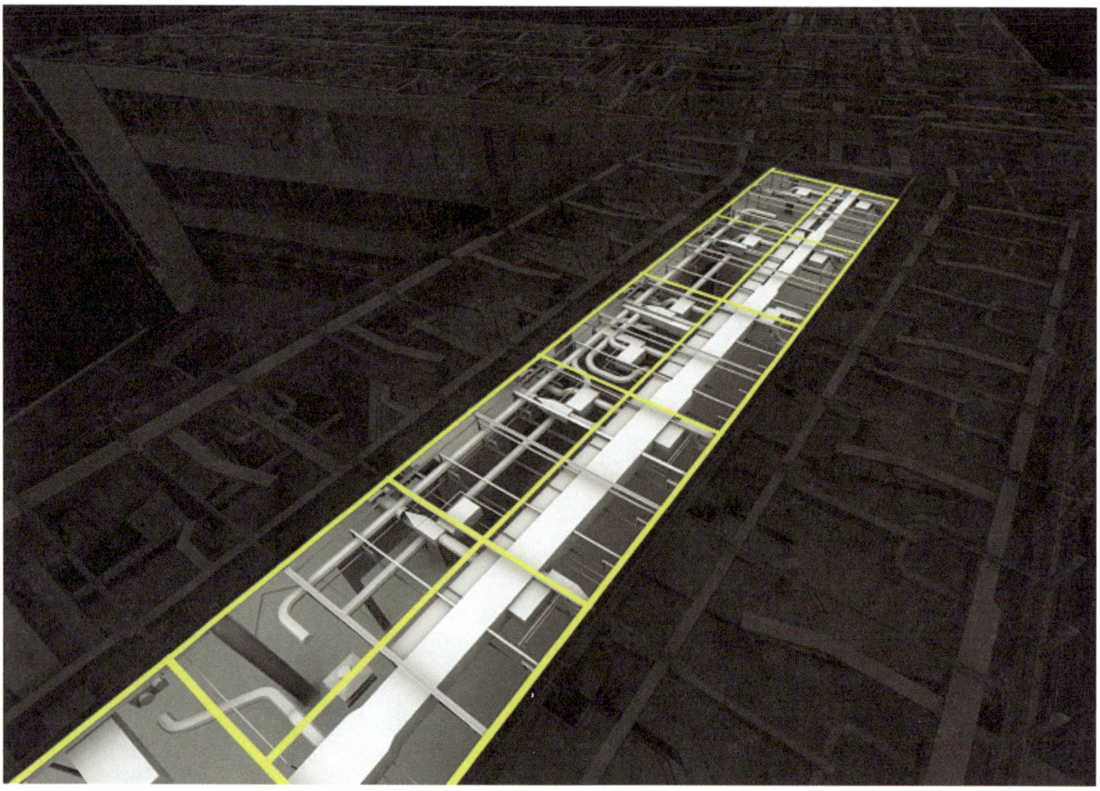

Figure 15.4 Location of the MEP racks over the 16-foot corridor. (Image courtesy of NBBJ.)

Figure 15.5 MEP racks in the prefab shop with the smoke-tight partition wall and MEP racks being lifted on the site. (Image courtesy of NBBJ.)

4. *The integrated MEP racks above corridors.* All the HVAC ducts and control boxes, gas, plumbing, sprinkler lines, electrical conduit, and cable trays were placed inside 210 MEP prefabricated racks. The 8- x 22-foot racks were designed to fit on top of the 16-foot wide corridors of the hospital wings (Figure 15.4). They were built in the same warehouse in which the inpatient rooms were prefabricated, joining the mechanical, electrical, plumbing, and drywall trades under one roof.

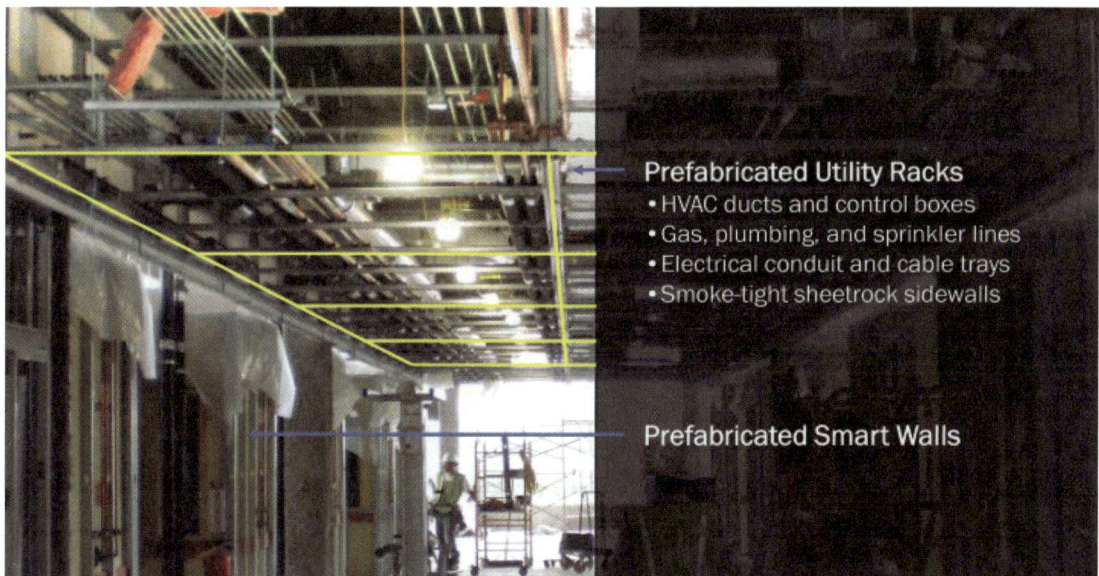

Figure 15.6 Prefabricated MEP racks placed at the construction site. (Image courtesy of NBBJ.)

Figure 15.7 The prefab bridge was installed in 3 days. (Image courtesy of NBBJ.)

The design of the racks served also to organize the systems that go above the ceiling in order to aid the maintenance of the facility. The rack frames also served as seismic restraint for all the MEP systems loaded in them. Every system that crossed from rack to rack needed its own coupling device. In this case they designed 1 foot of flexible pipe that was left loose at one end and was connected when they installed the racks on site.

Along one of the long sides of the racks a smoke-tight partition wall was placed to complete the corridor wall that separated the inpatient rooms (Figure 15.5). The racks were transported to the site in groups of two per flatbed truck, and it took 1 day to haul and place a full floor of racks and 1 to 1.5 weeks to position and secure the racks in their final position (Figure 15.6).

V. Temporary pedestrian footbridge. In order to build the new project three

Figure 15.8 Better ergonomic positions for construction workers. (Image courtesy of NBBJ.)

old buildings had to be demolished and a temporary footbridge had to be built to bypass the construction site. To construct a bridge via conventional delivery would have cost an estimated $2,100,000 and would have disrupted the entry to the hospital campus for 4 to 6 months. Instead the design/construction team developed a prefab solution with a company that builds passenger boarding bridges for airports. The prefab solution cost $980,000 to construct and was installed in 3 days at the site.

Prefabrication Performance Metrics

For Miami Valley Hospital NBBJ and Skanska Shook followed several performance metrics. Architect Ryan Hullinger from NBBJ says they noted improved construction quality, reduced errors, improved productivity, improved worker safety and ergonomics, better flexibility for the system, accelerated schedule, and significantly reduced waste. All the metal studs, ductwork, conduits, and pipes were ordered to length and less than one dumpster of waste was filled for the whole production of the prefabricated patient rooms and MEP racks. Hullinger says one example of the improved construction quality can be seen in the ducts and pipe systems, which in this prefab process were easy to inspect, test, clean, and cap before taking them to the site. These systems had fewer connections and were laid out to ease serviceability. The work at the prefab shop was more productive. For example, each plumber was installing 600 feet of pipe per day—a 300% increase from a conventional job site. Only 18 workers prefabricated the 180 patient rooms and 210 MEP racks. Architect Ryan Hullinger says that just by looking at those few items one began to see clearly a significant trend: fewer people, doing more work faster, in a safer environment, and all at a lower cost.

Neuroscience Institute at Riverside Methodist Hospital

Based on the experience presented above, NBBJ engaged in a second customized prefab endeavor when it was asked to be the architects for the Neuroscience Institute at Riverside Methodist Hospital in Columbus, Ohio. This hospital is a 400,000+ -square foot, nine-floors, and 224 private room

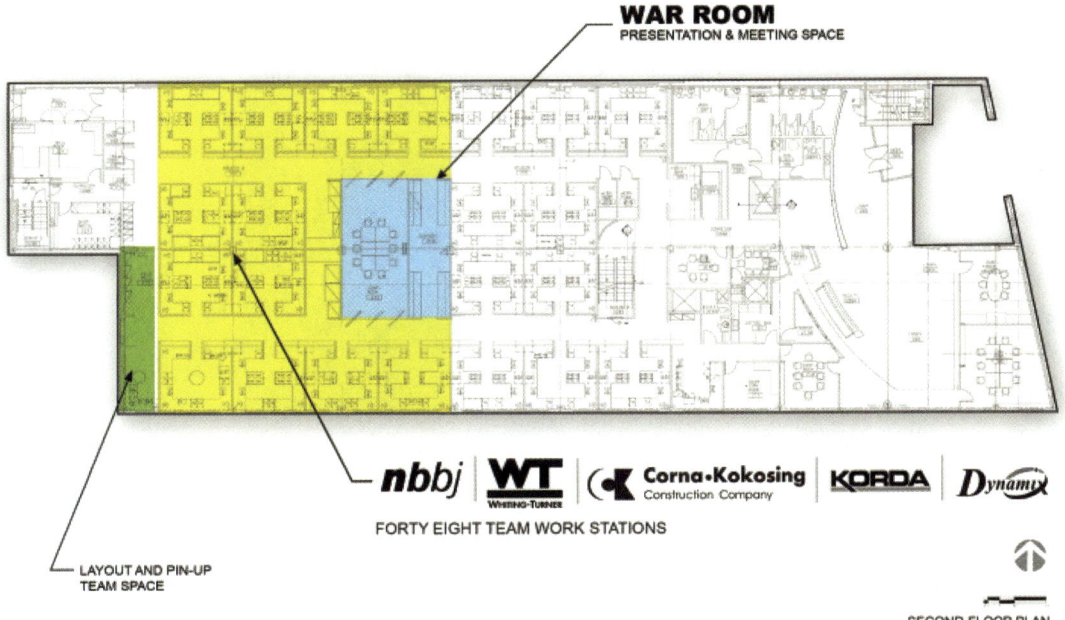

Figure 15.9 Floor plan of the colocation of the design/build team at NBBJ offices in Columbus, Ohio. (Image courtesy of NBBJ.)

neuroscience-oriented facility. This time NBBJ worked with Whiting-Turner Contracting Co. in a design-build contract.

Colocating of the Entire Design-Build Team in NBBJ Office

NBBJ Principal Tim Fishking says that one of the unique aspects of this second project was that the contractor occupied space within the architect's office in Columbus, Ohio, for 2 years, from the early programming phase all the way until the final construction documents. The MEP engineering team, interior designers, and landscape architects were also colocated in the same office at the beginning of the process. Early in the design development phase they also brought in design assist subcontractors for the MEP trades and for the exterior closure. The colocation of the entire design/build team used 48 workstations plus dedicated meeting and pin-up spaces. This space encouraged seamless communication and close collaboration between the design and construction teams. The contractors were invited to design critiques and charrettes with the client, which led the entire team to understand the building at a much higher degree than if everybody just looked at a set of drawings.

Just-In-Time Prefab Construction Schedule

Prefab was one of the main drivers of the process from day one, and the Miami Valley project was the model. However, in this project the construction schedule was built around prefabrication. By contrast, the Miami Valley project had a conventional construction schedule. Many of the prefab components in the previous project were built more quickly than anticipated and had to wait in the warehouse until the project site could receive them. In the Riverside project they decided to assemble, deliver, and install the modular components with a just-in-time (JIT) strategy. This

Figure 15.10   In the prefab shop for the Neuroscience Institute at Riverside Methodist Hospital in Columbus, Ohio, the construction team assembled all the MEP racks per floor following the curvature of the corridors in the construction site. (Image courtesy of NBBJ.)

Figure 15.11   MEP racks placed in the curved corridors of the Neuroscience Institute at Riverside Methodist Hospital in Columbus, Ohio. (Image courtesy of NBBJ.)

meant that there was a very precise construction sequence between the job site and the prefab warehouse. For example all the prefab components for the third floor were being built in the warehouse as the third floor was being built on site.

3-D and 4-D BIM Models

The entire design team worked from schematic design to construction documents using a single coordinated 3-D BIM model in Revit. That model went directly into Navisworks, which was used by the contractor team. Whiting-Turner Contracting Co. developed 4-D BIM construction sequencing models to diagram how the prefab warehouse would function and zone the production, storage, and loading of the components. The contractor also used BIM to do its cost-estimation processes.

Improvements in the Prefabrication of the Components

In the Riverside project the design of the inpatient organization, bathroom, walls, and MEP prefab components were similar to Miami Valley but what was different in this project was that the project followed a slight curvature in the floor plan. In the prefab warehouse there was a whole MEP rack production area that simulated the curvature of the corridors (Figures 15.10 and 15.11). There were several improvements in this new project. For example they added wheels to the MEP racks so they could be rolled to the truck and when they arrived to their floor at the jobsite. They improved the positioning of the MEP racks in the corridors by locating via Global Positioning System (GPS) ceiling tracks to which the prefabricated components were attached. Also the transportation of the prefab toilets to the jobsite was enhanced by building a lift carriage that eliminated the need for temporary hoist bracing every unit.

Conclusion

Both hospitals are among the ground-breaking examples of how design and construction teams are beginning to merge BIM, integrated practices, and high levels of prefabrication. What is unique about these two cases is that they are driven not by the economic efficiency of prefabrication but by the high level of customization that is needed in the health care industry today. Ryan Hullinger explains that in the case of NBBJ they usually do not begin the design with drawings but with 3-D models and with life-sized mock-up rooms in which health care teams use the spaces, perform simulations, and test their interaction with the equipment.

Hullinger said that they found that prefabrication was good for highly repetitive building components and highly complex building elements and an inpatient unit has both. He adds that today there is a significant movement in hospital administration and health care design in general to focus on standardization as a means of delivering better care. Standardization leads to error reduction and prevents staff disorientation. Customized prefabrication is an extension of that movement.

Chapter 16

Robotic Fabrication: ICD/ITKE Research Pavilion 2012

Achim Menges and Jan Knippers

Introduction

In November 2012, the Institute for Computational Design (ICD) and the Institute of Building Structures and Structural Design (ITKE) at the University of Stuttgart completed a research pavilion that is entirely robotically fabricated from carbon and glass fiber composites. This interdisciplinary project, conducted by architectural and engineering researchers of both institutes together with students of the faculty and in collaboration with biologists from the University of Tübingen, investigated the possible interrelation between biomimetic design strategies and novel processes of robotic production. The research focused on the material and morphological principles of arthropods' exoskeletons as a source of exploration for a new composite construction paradigm in architecture.

At the core of the project was the development of an innovative robotic fabrication process within the context of the building industry based on filament winding of carbon and glass fibers and the related computational design tools and simulation methods. A key aspect of the project was to transfer the fibrous morphology of the biological role model to fiber-reinforced composite materials, the anisotropy of which was integrated from the start into the computer-based design and simulation processes,

Figure 16.1 View of the finalized pavilion.

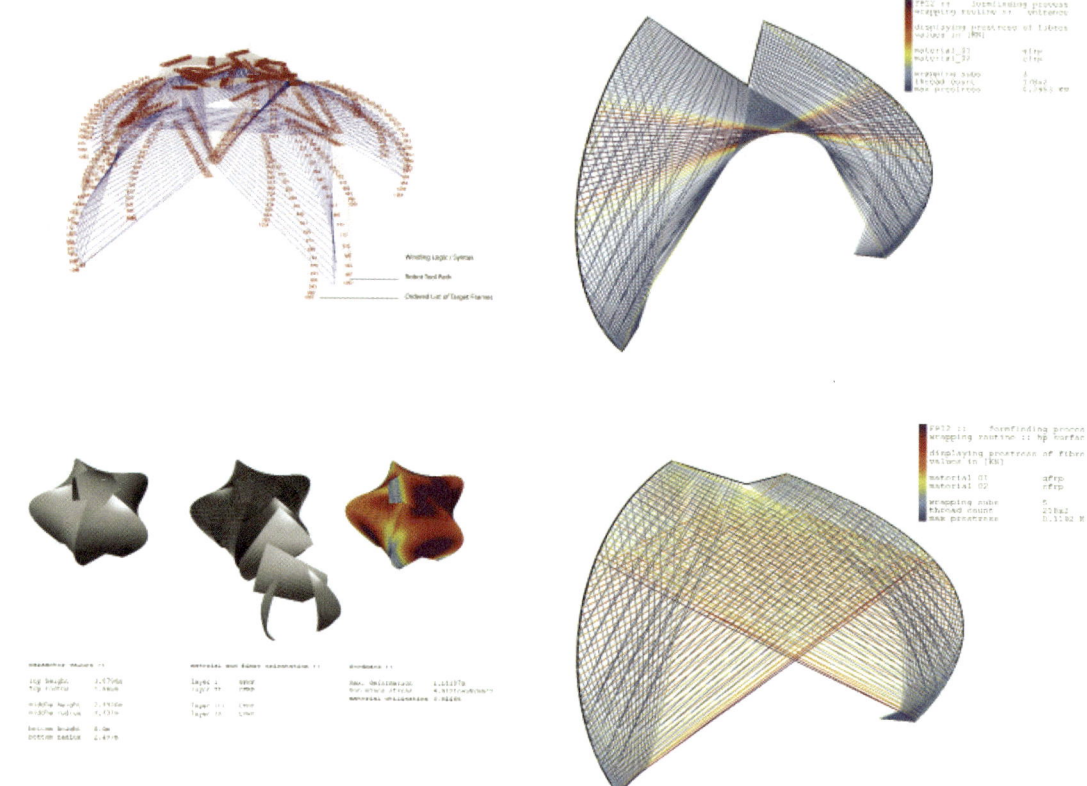

Figure 16.2 Computational simulations and optimizations.

thus leading to new tectonic possibilities in architecture. The integration of the form generation methods, the computational simulations, and robotic manufacturing specifically allowed the development of a high-performance structure: the pavilion requires only a shell thickness of 4 mm of composite laminate while spanning 8m (Figures 16.1–16.4, 16.7).

Biological Model

Following a bottom-up approach, a wide range of different subtypes of invertebrates were initially investigated with regard to the material anisotropy and functional morphology of arthropods. The observed biological principles were analyzed and abstracted in order to be subsequently transferred into viable design principles for architectural applications. The exoskeleton of the lobster (*Homarus americanus*) was analyzed in greater detail for its local material differentiation, which eventually served as the biological role model of the project.

The lobster's exoskeleton (the cuticle) consists of a soft part, the endocuticle, and a relatively hard layer, the exocuticle. The cuticle is a secretion product in which chitin fibrils are embedded in a protein matrix. The specific differentiation of the position and orientation of the fibers and related material properties respond to specific local requirements. The chitin fibers are incorporated in the matrix by forming individual unidirectional layers. In the areas where a nondirectional load transfer is required, such individual layers are laminated together in a spiral (helicoidal) arrangement. The resulting isotropic fiber structure allows a uniform load distribution in

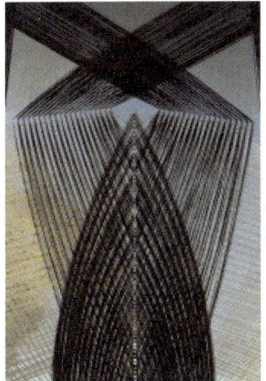   

Figure 16.3   Detailed views on fiber structure.

every direction. On the other hand, areas which are subject to directional stress distributions exhibit a unidirectional layer structure, displaying an anisotropic fiber assembly that is optimized for a directed load transfer. Due to this local material differentiation, the shell creates a highly adapted and efficient structure. The abstracted morphological principles of locally adapted fiber orientation constitute the basis for the computational form generation, material design, and manufacturing process of the pavilion.

Transfer of Biomimetic Design Principles

In collaboration with the biologists, the fiber orientation, fiber arrangement, and associated layer thickness and stiffness gradients in the exoskeleton of the lobster were carefully investigated. The high efficiency and functional variation of the cuticle are due to a specific combination of exoskeletal form, fiber orientation, and matrix. These principles were applied to the design of a robotically fabricated shell structure based on a fiber composite system in which the resin-saturated glass and carbon fibers were continuously laid by a robot, resulting in a compounded structure with custom fiber orientation.

In existing fiber placement techniques (e.g., in the aerospace industry or advanced sail production), the fibers are typically laid on a separately manufactured positive mold. Since the construction of a complete positive formwork is fairly unsuitable for the building industry, the project aimed to reduce the positive form to a minimum. As a consequence, the fibers were

Figure 16.4   Exterior and interior views.

laid on a temporary lightweight, linear steel frame with defined anchor points between which the fibers were tensioned. From the straight segments of the prestressed fibers, surfaces emerged, that resulted in the characteristic double-curved shape of the pavilion. In this way the hyperbolic paraboloid surfaces resulting from the first sequence of glass fiber winding serves as an integral mold for the subsequent carbon and glass fiber layers with their specific structural purposes and load-bearing properties. In other words, the pavilion itself establishes the positive formwork as part of the robotic fabrication sequence. Moreover, during the fabrication process it was possible to place the fibers so that their orientation is optimally aligned with the force flow in the skin of the pavilion. Fiber optic sensors, which continuously monitor the stress and strain variations, were also integrated in the structure. The project's concurrent consideration of shell geometry, fiber arrangement, and fabrication process leads to a novel synthesis of form, material, structure, and performance (Figures 16.5 and 16.6).

Through this high level of integration the fundamental properties of biological structures were transferred:

- Heterogeneity: six different filament winding sequences control the variation of the fiber layering and the fiber orientation of the individual layers at each point of the shell. They are designed to minimize material consumption while maximizing the stiffness of the structure resulting in significant material efficiency and a very lightweight structure.

- Hierarchy: the glass fibers are mainly used as a spatial partitioning element and serve as the formwork for the following layers, while the stiffer carbon fibers contribute primarily to the load transfer and the global stiffness of the system.

- Function integration: in addition to the structural carbon fibers for the load transfer and the glass fibers for the spatial articulation, functional fibers for illumination and structural monitoring can be integrated in the system.

Computational Design and Robotic Production

A prerequisite for the design, development, and realization of the project was a closed, digital information chain linking the project's model, finite element simulations, material testing, and robot control. Form finding, material, and structural design were directly integrated in the design process, whereby the complex interaction of form, material, structure, and fabrication technology could be used as an integral aspect of the biomimetic design methodology. The direct coupling of geometry and finite element simulations into computational models allowed the generation and comparative analysis of numerous variations. In parallel, the mechanical properties of the fiber composites determined by material testing were included in the process of form generation and material optimization. The optimization of the fiber and layer arrangement through a gradient-based method allowed the development of a highly efficient structure with minimal use of material.

The robotic fabrication of the research pavilion was performed on-site in a purpose-built, weatherproof manufacturing environment by a 6-axis robot coupled with an external seventh axis. Placed on a 2m high pedestal and reaching an overall working span and height of 4m, the robot placed

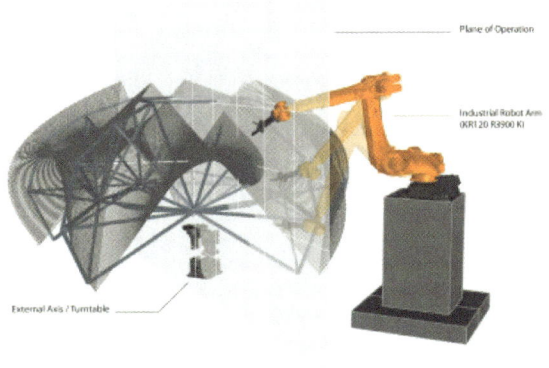
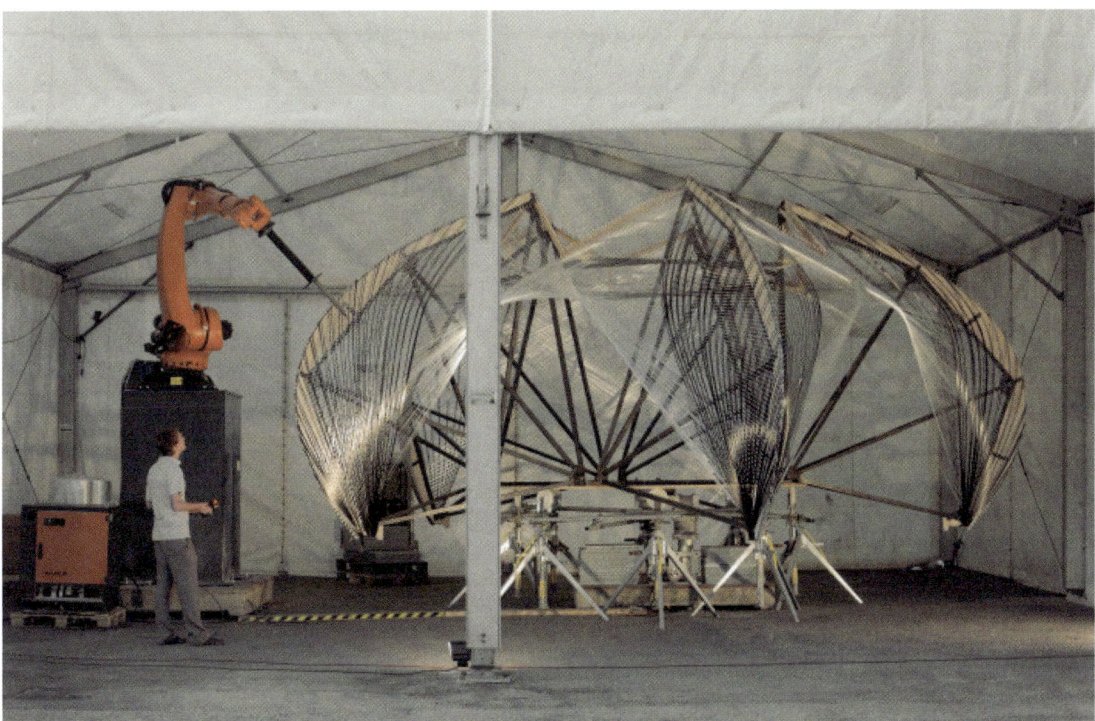

Figure 16.5 The robotic fabrication was conducted in a weather protected interior environment.

the fibers on the temporary steel frame, which was actuated in a circular movement by the robotically controlled turntable. As part of the fabrication process the fibers were saturated with resin while running through a resin bath directly prior to the robotic placement. This specific setup made it possible to achieve a structure of approximately 8.0 m in diameter and 3.5m in height by continuously winding more than 60 km of fiber rovings. The parametric definition of the winding motion paths in relation to the digital geometry model, the robotic motion planning including mathematical coupling with the external axis, as well as the generation of robot control code itself could be implemented in a custom-developed design and manufacturing integrated environment. After completion of the

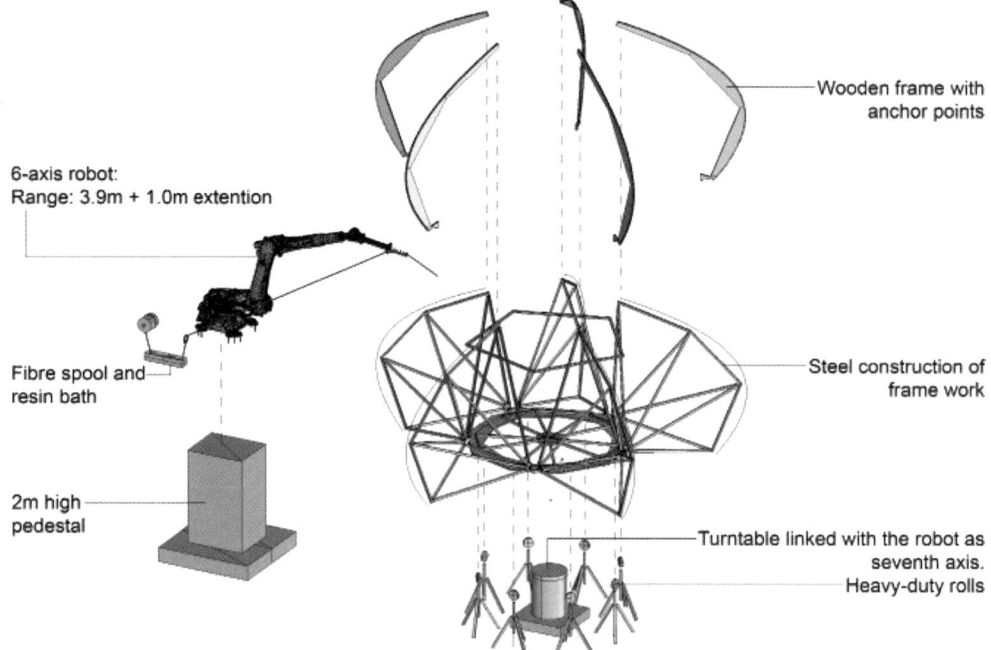

Figure 16.6 Diagram of the robotic production processes, equipment and tools.

robotic filament winding process and the subsequent tempering of the fiber-resin composite, the temporary steel frame could be disassembled and removed. The remaining extremely thin shell of just 4-mm thickness constitutes an automatically fabricated but locally differentiated structure.

The concurrent integration of the biomimetic principles of the lobster's cuticle and the logics of the newly developed robotic carbon and glass fiber filament winding within the computational design process enable a high level of structural performance and novel tectonic opportunities for architecture. Despite its considerable size and span, the semitransparent skin of the pavilion weighs less than 320 kg and reveals the system's structural logic through the spatial arrangement of the carbon and glass fibers. The synthesis of novel modes of computational and material design, digital simulation, and robotic fabrication allows both the exploration of a new repertoire of architectural possibilities and the development of extremely lightweight and materially efficient structures.

Project Data
Address: Keplerstr. 11-17, 70174 Stuttgart
Date of completion: November 2012
Footprint Area: 29 m²
Volume: 78 m³
Construction weight: 5.6 kg/m²
Material: Mixed laminate consisting of epoxy resin and 70% glass fibers + 30% carbon fibers
Project Team: Institute for Computational Design (ICD) - Prof. Achim Menges, Institute of Building Structures and Structural Design (ITKE) - Prof. Dr.-Ing. Jan Knippers
Concept Development: Manuel Schloz, Jakob Weigele
System Development & Realization: Sarah Haase, Markus Mittner,

Figure 16.7 View of the finalized pavilion.

Josephine Ross, Manuel Schloz, Jonas Unger, Simone Vielhuber, Franziska Weidemann, Jakob Weigele, Natthida Wiwatwicha with the support of Michael Preisack and Michael Tondera (Faculty of Architecture Workshop)
Scientific Development & Project Management: Riccardo La Magna (structural design), Steffen Reichert (detail design), Tobias Schwinn (robotic fabrication), Frédéric Waimer (fiber composite technology & structural design)
In Collaboration With: Institute of Evolution and Ecology, Department of Evolutionary Biology of Invertebrates. University of Tübingen - Prof. Oliver Betz. Centre for Applied Geoscience, Department of Invertebrates-Paleontology. University of Tübingen - Prof. James Nebelsick. ITV Denkendorf - Dr.-Ing. Markus Milwich
Main Sponsors: KUKA Roboter GmbH, Kompetenznetz Biomimetik, SGL Group, Momentive
More information: http://icd.uni-stuttgart.de/?tag=researchpavilion2012. http://www.itke.uni-stuttgart.de/entwicklung.php?id=30

Part IV
Emerging Automations

Chapter 17

Automating Design via Machine Learning Algorithms

Alfredo Andia

Introduction

Part II dealt with a large number of narratives from architects and engineers who are automating their processes. These cases are extraordinary and are today at the forefront of practice. The cases presented clearly show that in this post-parametric era algorithmic thinking has moved designers into making higher-level decisions about how to automate their design workflow.

This part will try to answer the following questions: What is next? Which computing methods could further automate significant parts of architectural and engineering design processes? By automation we do not mean the development of systems that imitate what humans do today but we imply the creation of new methods that can significantly augment and renovate contemporary design processes.

Limitations of Parametric Systems

In design theory in computer science parametric systems are considered the most primitive and archaic stage of artificial intelligence (AI). In a parametric system an expert, a programmer, has to manually code all the parameters. Anyone that has dealt with parametric environments knows that major workflows in design are an extremely hard, if not impossible, task-to code. The more factors or rules you include in a parametric model the exponentially more difficult it becomes to bond design associations.

Today a significant number of narratives about automation deal with the concept of parametric or ruled-based scripting. These narratives have been developed with some success in a number of areas. However, these endeavors are limited and are far from fully automating major architectural and engineering design workflows in the AEC industry. Those that optimistically advocate parametricism imply that the more parameters are programmed into a digital environment the more automated the design process will became. They project linearly that if they can code with parametric tools the design of a façade, a series of panels, or a detail today then it is just a matter of time before all types of design and construction information could be manually coded into a universal parametric model.

In this chapter, we put forward the idea that at present we are well into a second era of AI. In this new age, machine-learning algorithms can perform automated tasks by being trained from previous data. We suggest that these computing methods will be highly influential in the next generation of automation of design in the AEC industry.

Algorithms vs. Learning Algorithms

There is a difference between an algorithm and a learning algorithm. An algorithm is a set of instructions to perform a particular calculation or procedure. A learning algorithm is a set of instructions for a computer to learn from data and perform an action without the assistance of a human. Today learning algorithms that have been trained to detect traffic signs double the performance than that of humans. A whole generation of diverse products, including Internet search, automated translation, forecasting energy production, the driverless car, automated trading, drug design, and fraud detection are the results of learning algorithms.

With learning algorithms we are moving away from manually coding systems to designing systems that learn from experience. We present at the end of this chapter an example of how this could be possible between the framework of architectural design with a method that automatically generates programs, 2-D floor plans for residential buildings, and 3-D models in different architectural styles—100,000 iterations of 2-D floor plans in 35 seconds. We finally discuss why parametric is a limited paradigm for the future of design automation in the construction industry.

Computers as Autopoietic, Self-Organizing, and Self-Learning Systems

Terry Winograd and Fernando Flores wrote a book in 1987 called *Understanding Computers and Cognition: A New Foundation for Design* [1, 2]. Winograd is a highly influential AI professor at Stanford who was also the Ph.D. advisor of the doctoral thesis that was converted into Google. The book has proven to be deeply influential in computer design, however it is hardly known in architectural or engineering design circles. The book was written in a period in which AI was highly discredited.

The book works on developing a new understanding of what intelligence and cognition mean in the context of designing computer systems. The authors discredit the rationalistic approach used in AI during the 1980s. At that time, AI was heavily based on formal representations of intelligence such as rules, knowledge bases, and operations that describe intelligence in very narrow terms.

Instead the authors are inspired by the concept of autopoietic or self-organizing systems developed earlier by biologists Humberto Maturana and Francisco Varela [3]. These biologists argued that living organisms have extraordinary design intelligence because they self-organize by learning from their environments and they are not planned from the exterior.

Winograd and Flores criticize the old rationalistic software design because they create intelligent systems that are programmed by an outside coder, thus are not self-organizing or autopoietic. They concluded at the time that new approaches for designing computer systems were needed and put forward the notion of computer systems that somehow learn by themselves from their environment and self-organize. Winograd and Flores put forward the notion that the design of AI systems would come in three different stages of computer learning:

1. Parametric;
2. Machine learning;
3. General AI.

Parametric: First Stage of AI

In the first and most prehistoric stage of AI one can find parametric systems that via parameter adjustments or combinatorial search do basic intelligent operations. Winograd and Flores mention that after a short-lived peak in the 1950s and 1960s this type of work has been almost fully abandoned in the computer science community. Parametric allows for the coding of human reasoning; however, it always requires the hand of a coder that is expected to be able to observe all the potential steps of every condition of intelligent behavior.

Contemporary parametric endeavors in the design of buildings are good examples of the first age of AI in the architecture and engineering domain. This first age has allowed architects to make more explicit their design processes, but most of these systems are not self-organizing or autopoietic. These parametric systems can usually only tackle a few parameters. They are usually used in very particular evaluations but cannot fully automate large workflows of design processes.

Machine Learning: Second Stage of AI

A second level of AI impacts occurs when computers perform concept learning and concept formation. Here algorithms learn from data. These learning algorithms find patterns in data and build probability or predictive models for a specific job. Today learning algorithms work in a large array of tasks, they are usually very specific, and the techniques have spread rapidly in the computer science discipline.

Arthur Samuel, one of the early pioneers of AI, described machine learning as the "field of study that gives computers the ability to learn without being explicitly programmed" [4]. Automated learning systems learn from information using techniques such as artificial neural networks, decision tree learning, support vector machines, Bayesian networks, Boltzmann machines, and deep learning, among many others.

Machine learning algorithms can be classified into several types depending on the desired results and data available. They can be categorized as supervised learning, unsupervised learning, semisupervised learning, reinforcement learning, learning to learn, or transduction.

Today, learning algorithms are everywhere. They are automatically selecting companies for venture capital firms, and they are automating the discovery processes of many large practices in the legal community. Complex algorithms are already replacing engineers in certain tasks of chip design, writing sport news, Web articles for *Forbes Magazine*, grading English essays, developing patrol routes for the Los Angeles police, and are at the core of the IBM's Watson supercomputer that beat two former human champions of the TV game *Jeopardy!* after just 2 years of training.

Examples of Machine Learning Algorithms Outside the AEC Industry

An example of supervised learning is a driverless car that is trained how to drive by creating a neural network system that captures images of the road, 3-D laser data, and at the same time records the steering directions of human drivers. Once the system is trained the car will capture images and 3-D data of the road as the car moves, and the steering direction will be controlled by the neural network optimized results.

Another example of learning algorithms is automated translation similar to Google Translate. A translating system developed with parametric techniques will require that the programmer manually code all meanings and double meanings of words from one language to another. Just a single word can have a vast array of meanings, which makes translation by parametric means an impossible task. Machine learning algorithms do not understand text but if they are fed two texts in two different languages they will be able to parse the text and detect probability patterns such as that every time the word "one" appears in English then the word "uno" shows up in the Spanish text. Complementary algorithms can further aid the process by creating phrase tables to help assure the particular meaning of a word or sets of words. The more text is fed to the algorithm the higher the certainty that the predictive model will find the right translation.

Learning Algorithms in Architectural Design

Machine learning algorithms in the architectural design domain can evolve in several directions. In Chapter 10, Lars Junghans describes in detail the direction toward "automated building optimization algorithms." Another subject that has deserved significant attention in the architectural domain has been the automation of floor plan design in the initial stages of a project. A large number of attempts can be traced back to the 1970s, including the work of Per Galle [5], Bill Mitchell [6], shape grammars, heuristic optimizations, or contemporary 3-D parametric modeling and scripting. Most of these previous approaches are rule-based systems that adjust particular arrangements. But none are able to fully automate the design synthesis process from creating automatically a program, 2-D plans, or 3-D models.

The most advanced method in the area of automation in design synthesis can be found in the research led by Stanford professor Vladlen Koltun, whose work has been focused on visual computing and design synthesis using machine learning. These systems are not programmed with rules but are trained by feeding them a list of data sets such as shapes, program, size, and adjacencies found in the real world. The learning algorithms of this Stanford group have allowed for the automated generation of the architectural program, floor plans, sections, and 3-D models based on data feed from a book using machine learning techniques.

Automated Design for Residential Building

In 2010, Vladlen Koltun with Paul Merrell and Eric Schkufza guided the completion of a method for automatically generating the spatial design of residential buildings with a complete automated generation of architectural program, floor plan layouts, elevations, and 3-D models. The methodology used a Bayesian network. A Bayesian network is a type of statistical model that represents the probabilistic relationship between variables and conditional dependencies—a highly popular method in second generation types of AI endeavors.

This Bayesian network is trained on existing residential housing data found in the book *Essential House Plan Collection* by Hanley Wood [7]. From that data the network produces an architectural program without human intervention. From the generated building program the team uses a stochastic optimization to automatically generate sets of floor plans. From

Figure 17.1 The image illustrates the three-stage automation process of the method led by Stanford professor Vladen Koltun and Stanford researchers Paul Merrell and Eric Schkufza. Left: The architectural program with rooms and adjacencies are automatically created. Center: Floor plans are produced automatically from the previously generated programs. Right: 3-D models are generated from the 2-D plans. (Images from Merrell, Schkufza, and Koltun, 2010 [9].)

the plans entire 3-D buildings are generated in different styles that were also extracted from the book (Figure 17.1).

Automating Building Layout Design

The automation of floor plans has received intense interest since the 1960s. These endeavors have ended up in hundreds of methods that have resulted in numerous prototypes that have not truly automated the complete building layout process [8]. Most of those early endeavors programmed algorithms that tried to formalize design rules, criteria, and relationships manually, but none of them truly automated the process.

Koltun and his group at Stanford studied how three residential architecture firms developed building layouts in practice. They observed that the architects entered into a very time-consuming iterative process that include the construction of bubble diagrams, adjacencies, program lists, and multiple concept drawings that attempted to match the room requirements in different floors. In 2010 they noted that "real-world architectural programs have significant semantic structure." For example, "the presence of three or more bedrooms increases the likelihood of a separate dining room" [9].

They argue that these types of relationships are vast in the architectural domain. For them it is not clear how the implicit knowledge found in the architectural practice domain "can be represented with a hand-specified set of rules or with an ad-hoc optimization approach. A data-driven technique is therefore more appropriate for capturing semantic relationships in architectural programs" [9]. Thus, instead of using a ruled-based system this method trains a Bayesian network with data obtained from the book.

The automation of this building design methodology is a three-step process: (1) The generation of the architectural program, (2) the automatic generation of a set of floor plans, and (3) the automatic generation of a 3-D model. Figure 17.1 and also a short video of their 2010 published paper clearly shows the process followed by this method [10].

I. Automatically generating architectural programs with Bayesian networks. The objective of the first step of this method was the training of a Bayesian network with real-world architectural data. To nourish the system the team at Stanford manually coded architectural programs that were selected from

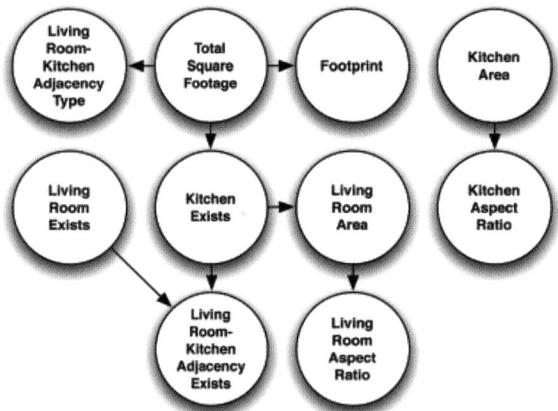

Feature	Domain
Total Square Footage	$\mathbb{Z}$
Footprint	$\mathbb{Z} \times \mathbb{Z}$
Room	{bed, bath, ...}
Per-room Area	$\mathbb{Z}$
Per-room Aspect Ratio	$\mathbb{Z} \times \mathbb{Z}$
Room to Room Adjacency	{true, false}
Room to Room Adjacency Type	{open-wall, door}

Figure 17.2 Illustration that summarizes a Bayesian network for hypothetical mountain cabins that contain a living room and an optional kitchen. The table on the top shows some the features coded from real-world data. The bubble diagram at the bottom represents the Bayesian network distribution of the features used to define the architectural program. (Images from Merrell, Schkufza, and Koltun, 2010 [9].)

120 projects from a popular book about residential design [7]. Features from every room of these 120 residential cases were recorded. The features tabulated included the program type, square footage, aspect ratio, adjacency, and other aspects that traditionally mediate the relationship among the rooms such as doors or open-wall connections. An example is shown in Figure 17.2.

The network-structured learning discovered a large number of relationships that were present in the data. For example, a room type such as a bedroom is an excellent forecaster of the room's size and aspect ratio. From this data the Bayesian networks can generate specific programs after 10, 100, and 1,000 iterations (Figure 17.3).

II. Automatically generating 100,000 floor plan iterations in 35 seconds. The second phase involves turning the architectural programs generated in the first phase into entire floor plans for each floor. Several techniques are used to align walls and rooms. Figure 17.4 illustrates several ill-formed floor plans that do not comply with a set of specific terms such as accessibility, area, aspect ratio, or shape that were introduced to improve the quality of the floor plans. Figure 17.5 shows the floor optimization process from 200 to 100,000 iterations that "took 35 seconds using an Intel Core i7 clocked at 3.2GHz" [9], similar to a typical processor found in a desktop or laptop computer in 2014.

III. Automatically generating 3-D models. From the building layouts generated in the second phase the team created 3-D models in different styles based on style templates that list the "geometric and material properties of every building element: windows, doors, wall segments,

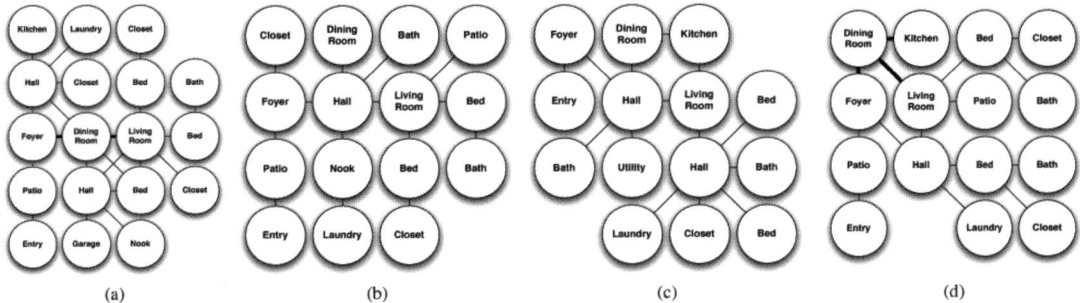

Figure 17.3 Illustration showing the architectural programs in bubble diagrams format that were generated from the Bayesian network after 10 iterations (a), 100 iterations (b), and 1,000 iterations (c). Diagram (d) was made by an architect for comparison. (Images from Merrell, Schkufza, and Koltun, 2010 [9].)

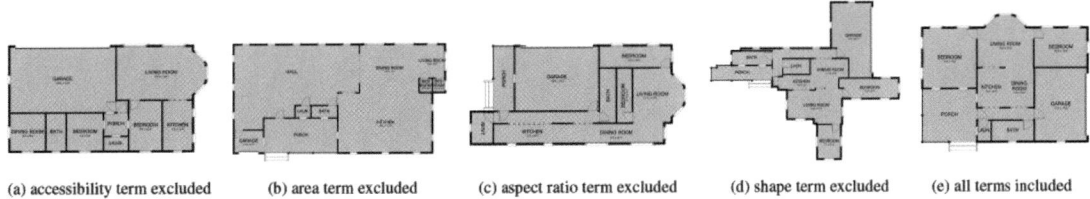

(a) accessibility term excluded (b) area term excluded (c) aspect ratio term excluded (d) shape term excluded (e) all terms included

Figure 17.4 Ill-formed floor plans generated by the learning algorithm. (Images from Merrell, Schkufza, and Koltun, 2010 [9].)

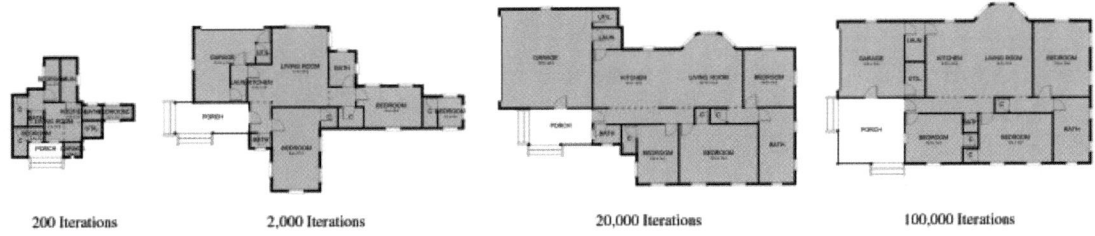

200 Iterations 2,000 Iterations 20,000 Iterations 100,000 Iterations

Figure 17.5 Images from the floor plan optimization. (Images from Merrell, Schkufza, and Koltun, 2010 [9].)

gables, stairs, roofs, and patio poles and banisters" [9]. Figure 17.6 illustrates four 3-D models in the cottage, Italianate, Tudor, and Craftsman styles that were automatically generated for the same 2-D building layout. The styles were extracted from the Hanley Wood book [7].

The work directed by Professor Koltun at Stanford is a highly sophisticated machine learning methodology applicable to the architectural domain. A more sophisticated type of method can be developed by training the machine learning algorithms to become skilled at styles or design concepts found in a diverse population of work and designers. Style, in a creative realm, is a higher-level semantic problem found in design. Initial ground has been developed in drawing and painting learning algorithms [11] and will certainly continue to evolve in the creative domains.

Machine Learning Hardware: Neuromorphic Processors

Most of the explosive advances in the machine learning field in the past decade has occurred in software design. However, the hardware of computers has remained very much related to the digital computer

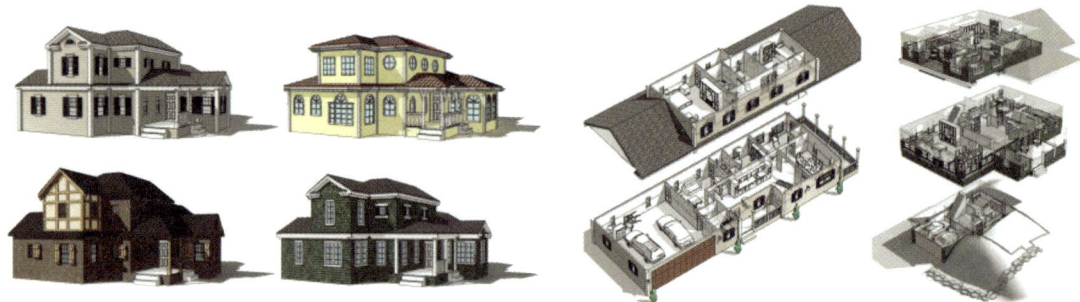

Figure 17.6 An example of the 3-D models generated from a single automated floor plan design. The architectural styles were obtained from the styles of residential houses that were in the book in which this project was based. (Images from Merrell, Schkufza, and Koltun, 2010 [9].)

architecture that emerged from the mathematician John Von Neumann around 1945. In very simplistic terms the Von Neumann model is a device that can calculate at very fast speed using strings of 1s and 0s. Today's machine learning algorithms usually have to represent, evaluate, and optimize data in a calculable format that a Von Neumann computer can handle. Thus, a significant part of machine learning efforts to date has been based on statistical-oriented algorithms that employ brute force, using massive computer power to perform a colossal number of calculations on a huge number of data sets.

Biological organisms are much more efficient in developing intelligent behavior than contemporary computers. There is an emerging field in hardware design that is developing neuromorphic processors that allow us to create a machine that truly learns by experience. Neuromorphic computer processors have an architecture that imitates the neurobiological design present in the nervous system of organisms. These electronic circuits are connected by wires, and their overall design mimics the morphology of neurons and biological synapses. Initial prototypes such as the ones sponsored by the DARPA SyNAPSE program at IBM and HRL Laboratories, Neurogrid at Stanford [12], and several projects under the sponsorship of the Human Brain Project (HBP) in Europe show early glimpses of computable brain simulation for tasks such as visual recognition, edge detection, music identification, pattern recognition, color identification, or smell. For example, a system of 120 neurons from HRL lab learned how to play the game Pong after five rounds of playing with the paddle and sensing the ball from the game. The system was not programmed; it only received feedback via rewards for a good job or punishment for a failure.

Neuromorphic systems will be complementary to today's computers whose capabilities continue to grow explosively. Several companies are announcing the commercial release of neuromorphic systems in 2014. They represent the next generation of technology that can move us closer to a third stage of AI systems.

General AI: Third Stage of AI

A third generation of AI will emerge when a device is no longer programmed but evolves and develops primarily by learning and can produce other machines even more intelligent than itself. Computer power and technology today is far from achieving the third stage of AI. But according to authors like Ray Kurzweil, now head of the Google mind project, by 2029 computer power will allow us to reverse engineer the human brain, which will be a significant advancement to create computers that learn by themselves [13]. Others like Jeff Hawkins, cofounder of the AI company Numenta, which has created some of the most sophisticated learning algorithms that operate like the neurons in the neocortex part of the brain, say they are very doubtful that this third stage will ever be achieved. In the meantime, at least for the next two decades, we are bound to observe the explosion of machine learning algorithms and hardware that are affecting many domains and industries.

Conclusion

The intention of this chapter was to place the automation efforts of the architecture and engineering community in context with the contemporary discourses of design theory in computer science today. We have put forth concrete examples about how machine learning is entering the domain of design. Machine learning algorithms are also entering into many others aspects in the construction industry such as BAS, procurement, and sensors systems.

Still there are doubts whether learning algorithms and neuromorphic processors could scale up to truly imitate the sophistication of biological systems. The subjects of AI, machine learning, and deep learning will continue to expand as computer power grows exponentially. There will be big AI booms and also major busts along the way. The second generation of AI learning algorithms are transforming many aspects in many industries and as illustrated in this chapter will also impact the construction industry in this post-parametric period. The major limitation is that there is very limited expert knowledge in learning algorithms available in the design community, which is the reason for the narrow use of this method in the AEC industry today.

References

[1] Terry Winograd is a Computer Science Professor at Stanford and today a highly influential figure in the Computer Science Community worldwide and well known in media in part because he was the advisor of the Ph.D. thesis of Larry Page, which became the Google search engine algorithm. Fernando Flores is a Chilean engineer who led the Cybersyn project with British scientist Stanford Beer from 1971–1973. Cybersyn was a network system that allowed the Chilean government to receive data from the state run enterprises and could monitor indicators and allowed statistical modeling analysis in real-time. The whole pioneering system was run using Telex machines.

[2] Winograd, Terry, and Fernando Flores. *Understanding Computers and Cognition: A New Foundation for Design.* Norwood, NJ: Ablex Pub., 1986.

[3] Maturana, H. R., and F. Varela. *De maquinas y seres vivos.* Santiago: Editorial Universitaria Santiago, 1973.

[4] Ratner, Bruce. "A comparison of two popular machine learning methods." *Mine Tech*, 2000.

[5] Galle, Per. "An algorithm for exhaustive generation of building floor

plans." *Communications of the ACM*, 24(12), 1981: 813–825.

[6] Mitchell, William J. *The Logic of Architecture: Design, Computation, and Cognition.* MIT Press, 1990.

[7] Wood, Hanley. *Essential House Plan Collection: 1500 Best Selling Home Plans.* Hanley Wood: Washington, DC, 2007.

[8] Liggett, Robin S. "Automated facilities layout: past, present and future. *Automation in Construction*, 9, 2000, 197–215.

[9] Merrell, Paul, Eric Schkufza, and Vladlen Koltun. "Computer-generated residential building layouts." *ACM Transactions on Graphics (TOG)*, 29(6), 2010, Article No. 181.

[10] Merrell, Paul, Eric Schkufza, and Vladlen Koltun. "Computer-Generated Residential Building Layouts." video, 13 Jan. 2014: http://vladlen.info/publications/computer-generated-residential-building-layouts/.

[11] Lindemeier, Thomas, S. Pirk, and O. Deussen. "Image stylization with a painting machine using semantic hints." *Computers & Graphics,* 37(5), 293–301.

[12] Choudhary, Swadesh, S. Sloan, S. Fok, A. Neckar, E. Trautmann, P. Gao, T. Stewart, C. Eliasmith, and K. Boahen, "Silicon Neurons that Compute." *Artificial Neural Networks and Machine Learning–ICANN 2012*, Springer, Heidelberg, 2012, 121–128.

[13] Kurzweil, Ray. *The Singularity is Near: When Humans Transcend Biology.* Penguin, 2005.

Chapter 18

Automating Construction via n-D Digital Manufacturing

Alfredo Andia

Introduction

In Part III, we presented a large number of extraordinary cases of prefabrication, modularization, fabrication, and manufacturing in construction. These are surprisingly big innovations in a construction industry that has not received major disruptive news in the past 60 years. An increasing number of prefab endeavors today are moving significant parts of their site labor into factories. These efforts are just steps away from more advanced factory automation narratives such as robotics, warehouse mechanization, and manufacturing computerization.

In this chapter, I argue that the prefabrication and factory automation narratives in the construction industry, although extraordinary, are still narrow. Information technology will "disrupt" or "hack" the construction industry in a parallel way. The current 3-D digital manufacturing technology will eventually evolve into designed materials, programmable matter, and intelligent apparatuses that will offer different alternatives to alter our analog world.

At the Dawn of Three New Manufacturing Eras

3-D manufacturing today encompasses automated additive technologies such as 3-D printing and subtractive technologies like computer-controlled machines (CNC), laser and water cutters, as well as automated assembly technologies such as industrial robots, quadricopters, and other programmable equipment.

The first thing to understand about computer-controlled 3-D manufacturing and 3-D printing is that these technologies are not going away. On the contrary, they are growing exponentially in many different directions in a very short time frame. What we are witnessing today is just the first generation of digital manufacturing. Computer-controlled manufacturing will be transforming in the coming years in three different distinct stages:

1. *3-D digitally controlled manufacturing:* Today we are at the maturing stages of the first era of digital manufacturing in which 3-D printers, CNC, laser cutters, robotic arms, and a whole array of digitally automated machines can print, cut, and carve any shape on industrialized materials.

2. *4-D digitally controlled manufacturing:* In a second period sophisticated multimaterial printers and nanotechnology will allow us to create new materials that can be designed and fabricated at a microscopic and atomic level. These new artificial materials will perform more efficiently than raw industrialized materials and will have behaviors that cannot be found in natural materials.

3. *n-D digitally controlled manufacturing:* In a third period there will be the emergence of programmable matter, synthetic biology, and evolutionary robotics; it will be an epoch in which we can manufacture digitally

transformable matter that can be programmed at will. We will also witness the emergence of apparatuses that can self-design and self-manufacture at different scales.

3-D Manufacturing: Making Any Form in 3-D

In the past few years a whole array of digitally controlled 3-D manufacturing apparatuses such as additive printing, computer-controlled laser cutters, robotic arms, and CNC machines have become very visible in mainstream media due in large part to the dramatic drop in the prices of entry-level machines. Inexpensive 3-D printers allow anyone to make prosthetic hands, guns, toys, lamps, music records, and all types of objects or items.

Increasingly, every day there seems to be a new type of 3-D printer machine arriving. There is food-printing: from chocolate machines to 3-D pizza printers. There is bioprinting, in which one can print with living cells. Skin, ears, organs, and even bone replacements have been already bioprinted. There is even a do-it-yourself bioprinter for $150 made by the amateur bio-lab collective BioCurious.

3-D Manufacturing: Large-Scale Digital Manufacturing

While small 3-D manufacturing technologies at the consumer level are getting cheaper and cheaper, at the higher end they are getting larger and larger. There are giant 5-axis CNC such as the HSM-Modal by EEW Protec that is able to mill in a 1-to-1 scale anything from cars to ships, with a workspace area of more than 5,300 square meters.

Large-format 3-D printers like those produced by companies such as Voxeljet can print up to 4.0 m x 2.0 m x 1.0 m DUS Architects in the Netherlands created a large and movable PLA 3-D printing pavilion called The KamerMaker. That printer can print interiors of 2.0 m x 2.0 m x 3.5 m

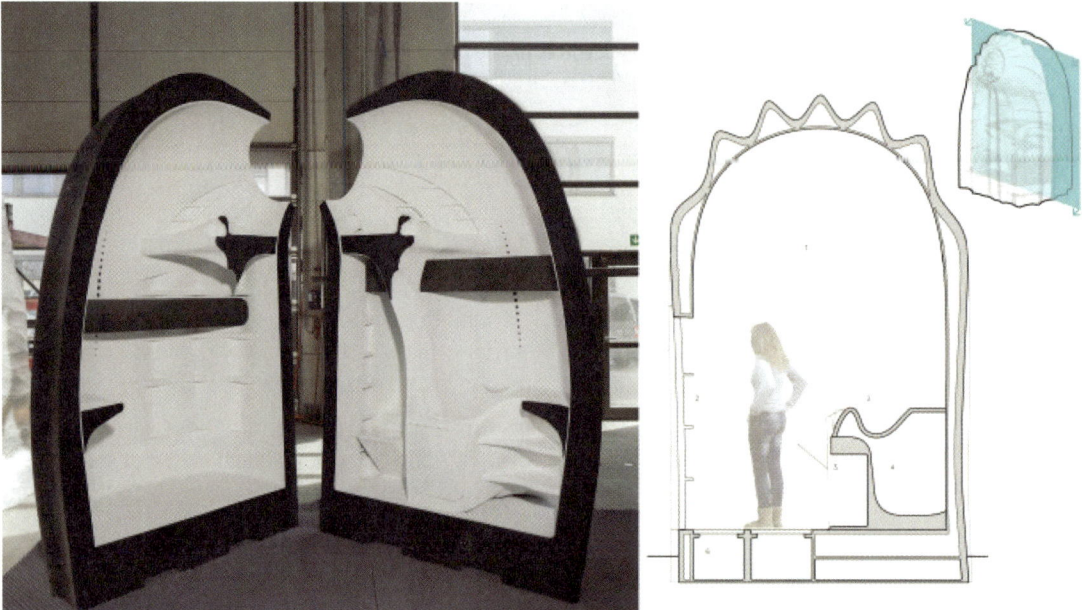

Figure 18.1 3-D printed micro-house of approximately 50-square-foot created by a studio lead by professor Peter Ebner at the 3M future lab at UCLA. (Drawings by students from 3M future lab at UCLA and pictures by Florian Holzherr.)

in height. D-Shape, perhaps the most famous large 3-D printer in media, has a rigid base of 6m x 6m, and the vertical columns can be extended to a height of 9 to12m. D-Shape prints, using an inorganic binder and sand or mineral dust. A gazebo sculptural piece of approximately 3m x 3m x 3m printed by this 3-D printer had a material cost of just £60 (approximately $100) for sand and binder.

D-Shape also won a competition in 2013 sponsored by the New York City Economic Development Corporation to find innovative ways to fix thousands of piles that support piers and bulkheads around the Manhattan riverfront. D-Shape winning proposal includes underwater 3-D scans of the damaged piles. The scanned information is used by a generative algorithm to produce the most efficient design to strengthen each pile. Then the 3-D printed column is packed in an inflatable raft that is moved into the water and slowly deflated as a team of divers place the pile reinforcement in the right place. D-Shape has already fabricated underwater reefs in the Persian Gulf and has also tested 3-D printing lunar bases for the European Space Administration (ESA) with replicated moon dust.

We are at the end of a 3-D digital manufacturing era today. All our major inventions in additive and subtractive technologies alter material extremely accurately at the 3-D level. But most of these technologies are limited because they can just transform industrialized matter at the human scale, which is still a highly intensive labor process.

4-D Manufacturing: Printing New Materials

A more disruptive era of digital manufacturing will arrive when digital manufacturing advances to produce new materials. A newer generation of 3-D microprinters, such as the one being developed by German company Nanoscribe, can print patterns of materials at a very small scale—even 30 nm. The properties of these new microscopically designed and microprinted materials cannot be found in nature. 4-D printing allows 3-D material to be programmed and self-assembled into different shapes and patterns to perform in ways no material can perform today.

We all have discovered that when we stretch a rubber band we can see that the rubber band gets thinner as we stretch it. That is called the Poisson effect or Poisson ratio. Hod Lipson from Cornell University has printed a material at the microscopic level that had a negative Poisson ratio. That means that the material of a rubber band becomes thicker as it gets stretched [1]. This peculiar material that cannot be found in nature is called Auxetic. Auxetic materials cannot be manufactured with traditional methods but can be printed on demand by high-resolution multimaterial printers. Auxetic materials can be used in car bumpers that when hit can direct the force of the crash to other parts of the vehicle—protecting the passengers. Multimaterials printed in different patterns can achieve more optimal behaviors than traditional materials in the construction industry.

The advent of new multimaterials printers also implies that the designer and the design process will also have to change. Traditional design processes that use CAD or BIM are usually unable to plan for material performance at the macroscopic and microscopic level. Design automation will have to play an increasingly important role in design synthesis for the construction elements that use multimaterials. Jonathan Hiller and Hod Lipson from the Creative Machine Lab at Cornell University have presented

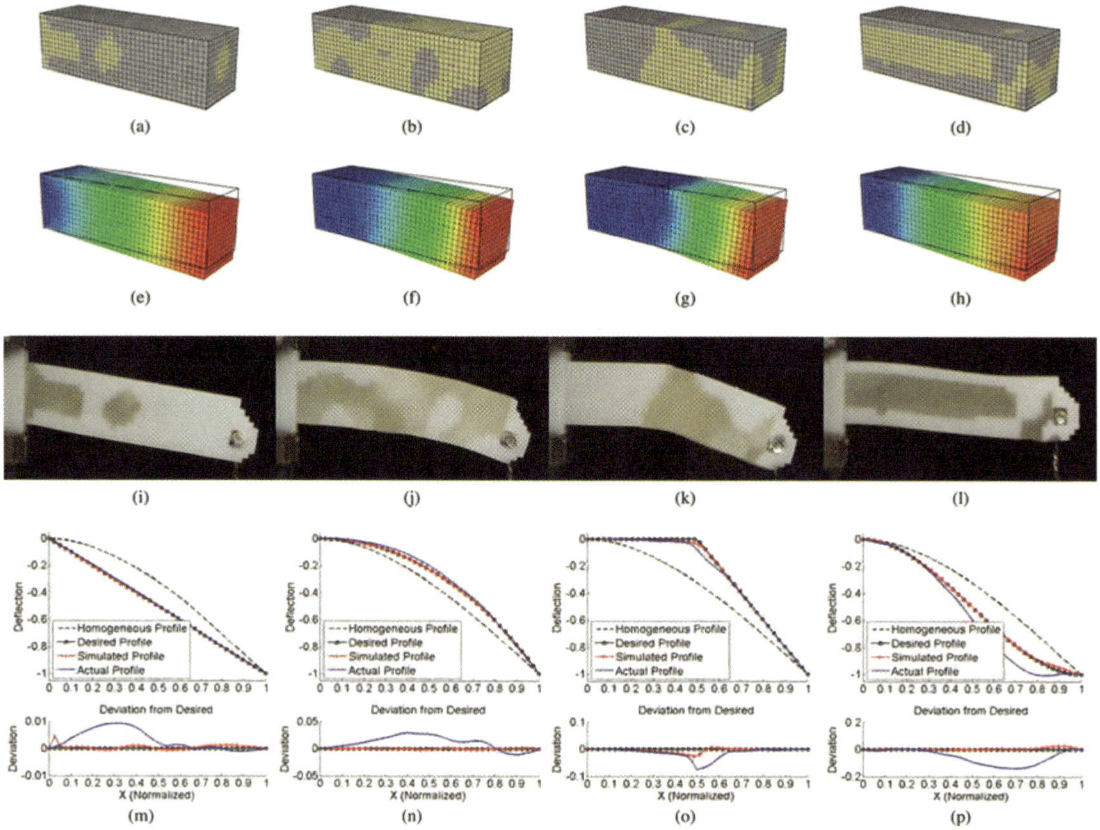

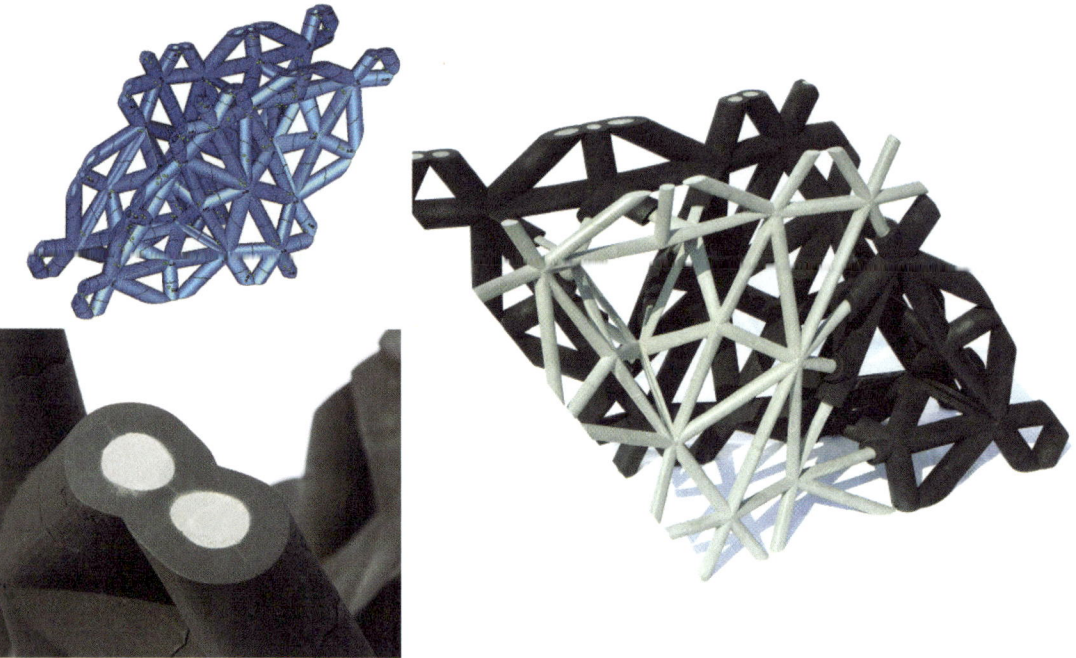

Figure 18.2 The material distribution of these 3-D printed cantilever beams was designed and fabricated automatically using an evolutionary algorithm [3]. (Image courtesy of Hod Lipson, Cornell University, Creative Machine Lab.)

Figure 18.3 EZCT Architecture & Design Research from Paris developed a recursive lattice structure project using ultra-high performance fiber-reinforced concrete. The highly complex biologically-inspired structure was casted in 3-D printed sand molds. (Image courtesy of EZCT Architecture & Design Research.)

a genetic algorithm (GA) method that is used to optimize the geometry of a multimaterial cantilever beam [2].

These authors state that the GAs are "suitable for designing the complex multi-material objects that have recently become possible to fabricate using additive manufacturing techniques… Instead of designing an object using traditional CAD programs, genetic algorithms allow an engineer to simply set high-level goals to be fulfilled and the blueprint is autonomously generated" [2].

4-D Manufacturing: Adaptive Materials

Skylar Tibbits, at MIT's Self-Assembly Lab, is working on 3-D printed pipes that work with adaptive materials that expand, contract, or undulate to change the capacity or flow rate of metropolitan water infrastructures. These pipes made with 3-D printed materials are programmed to transport liquid more like human veins transport blood, which is very different from the water infrastructure we have in our cities today full of pipes, pumps, and valves forcing water at constant rates.

4-D Manufacturing: Nanotechnology.

Nanotechnology and nanomaterials are part of another generation of 4-D manufacturing technologies that allow the manipulation of materials at an even smaller scale than 4-D printers, usually at scales lower than 100 nm—a human hair is around 80,000 to 100,000 nm in diameter. Nanotechnology is at various stages of maturity in the construction industry, ranging from experimental projects to available products [4]. Currently it is playing an increasingly important role in improving the basic properties of traditional construction materials. Today, nanotech particles are being mixed with traditional construction materials to improve basic properties. These benefits range from making concrete, steel, and wood stronger, to improving thermal insulation in panels, increasing solar cell efficiencies, waterproof coatings, self-cleaning and air-purifying paintings, and even making fire-protective glass and concrete.

In contrast to 3-D manufacturing, 4-D manufacturing is allowing us to manipulate materials at the microscopic, atomic, and molecular level—at an ultrasmall scale. 4-D printing and nanotechnology are still young but they are platform technologies that are growing very quickly and are one more step toward digitally based n-D manufacturing.

n-D Manufacturing: Programmable Matter

One of the most advanced metaphors in the future of computer-based manufacturing systems is called programmable matter. In this third phase we no longer just print inanimate objects but can manufacture matter that can be programmed. Matter that has the ability to change its shape, density, color, conductivity, and so forth. There are many approaches but perhaps the most common image about programmable matter is that it is made of minuscule 3-D pixels. These tiny 3-D pixels can assemble and disassemble as they are activated by a computer. There are several international projects such as Claytronics (Carnegie Mellon and Intel) and Robot Pebbles (CSAIL at MIT), but these systems at this time are very far from becoming a workable reality in the near future. Similarly, the most advanced visions in nanotechnology plan methods that can grow, assemble, and modify matter at the atomic and molecular levels.

n-D Manufacturing: Self-Made Robots

Another advanced paradigm in 3-D printing is evolutionary robotics, apparatuses that self-design and self-manufacture. Today, we design our material machines (including 3-D printing machines) manually. Robots and 3-D printing machines require large teams of design engineers, other sophisticated machines to produce them, and laborious processes of trial and error. Evolutionary robotics methodology is an emerging design methodology in 3-D printing and robotic communities. Its purpose is to create robots and/or machines (including 3-D printers) that could self-design, learn, and evolve their configurations over time, all by themselves.

The Symbrion project in Europe and the Evolutionary Robotics project at the Creative Machine Lab at Cornell are among the many research centers around the world that have built some of the earliest self-aware and self-replicating robots [5]. These self-made robots evolve first in a virtual 3-D environment that is full of robot parts and rules for assembly. The robots in the virtual environment use algorithms to test different populations of machines for a detailed job. The populations of candidate solutions are constantly modified to match the fitness function. They evolve from a selective set of forces, like organisms in nature.

The evolutionary robotics field is just at the very early stages of development but could potentially create any type of machine, including specific robots that work at the microscopic and nanometer scales. Creating machines to evolve without a human designer permits surprising results. Although the self-made machines created by the field of evolutionary robotics are still simple, they show another dimension of processes of building our analog world will be transformed in the near future into manufacturing controlled by computation.

n-D Manufacturing: Synthetic Biology

A parallel vision of programmable matter is rising in the growing field of synthetic biology. In synthetic biology living systems are engineered bottom-up to produce new materials and devices. We are at the early ages of this new paradigm but design firms such as The Living from New York, ecoLogicStudio from London, and Faulders Studio in Oakland have been engaging in creating new types of structures with biological processes. In 2014 The Living unveiled the Hy-Fi project at MoMA PS1 Young Architect program in New York. A 40-foot-tall tower was made of biomaterial bricks that were grown locally from mushroom mycelium and agricultural waste with almost no carbon emissions. A more advanced version of synthetic biology would allow the creation of environments that are made of lifeforms from redesigned or completely new biological entities.

Conclusion

Computer-controlled manufacturing will occur in three different phases. In this chapter these stages are called 3-D, 4-D, and n-D digitally controlled manufacturing. In the first period designers can 3-D print or 3-D manufacture any form. Design and construction workflows will not be

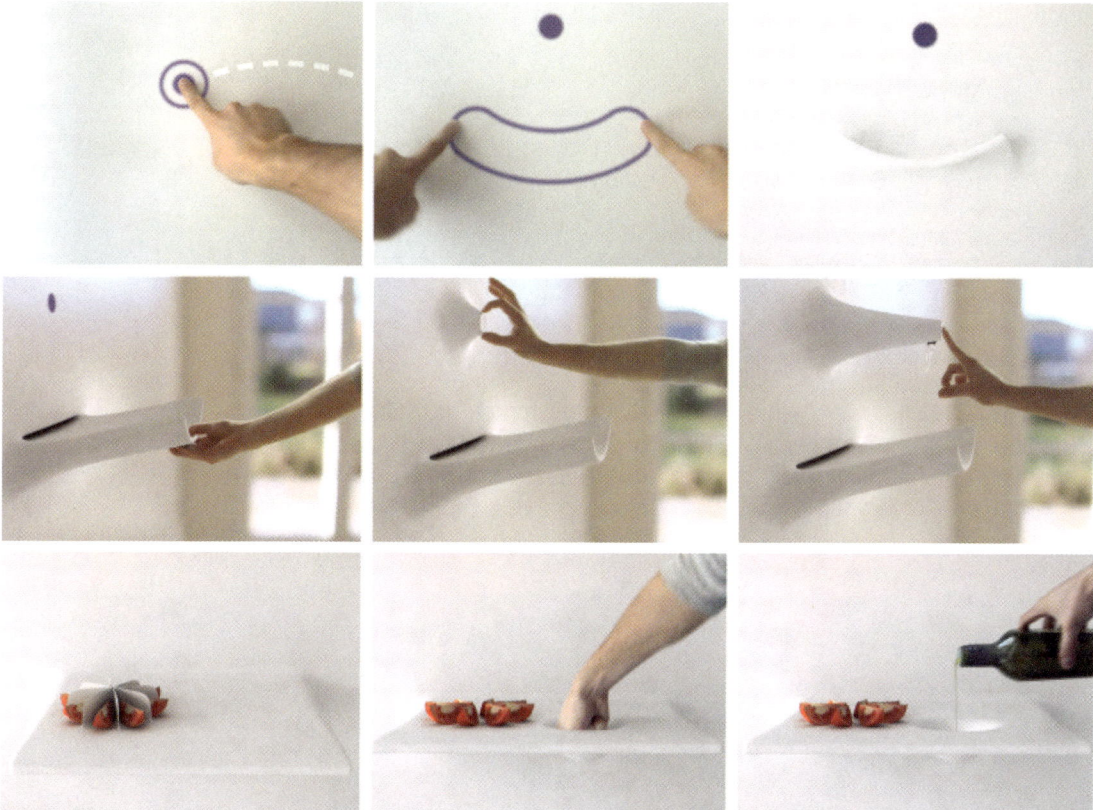

Figure 18.4 The "living kitchen" is a concept project developed by designer Michaël Harboun. The project explores a user interacting with programmable matter's ability to constantly change shape in a kitchen environment. In this prototype kitchen elements such as faucets, sinks, and cutting tables appear and disappear interactively as the user prepares a meal. The work is inspired on the Claytronics technology being developed at Carnegie Mellon University. (Image Courtesy of Michaël Harboun.)

altered significantly in this period. Most of the 3-D digital manufacturing technology to date just cuts, adds, or carves traditional materials at the human scale. For some observers the future of digital manufacturing is in the macro scale, with gigantic 3-D printers, large CNC systems, and powerful and cheap robotic machines. However, in this chapter we argue that a much more powerful stage of digital manufacturing will occur when we are able to manipulate matter at the micro scale.

A second period of digital manufacturing will emerge when technologies such as 4-D multimaterial printers and nanotechnology mature to alter materials at a very small scale. In this second age we will be able to digitally manufacture materials that cannot be found in nature. Further in the future, a final third period of multidimensional or n-D digital manufacturing will emerge as materials can be programmed, bio-engineered, and self-replicated at even a nanometer scale.

In the next 30 years we will follow a clear process of transformation in the production of our analog world. The shift toward digitally and biologically controlled manufacturing will also mean a change in the traditional role of designers and contractors. This will mean a gradual shift from manually based design and construction to increasingly automated processes that not only manipulate traditional materials but also design the properties of materials even at the macroscopic, atomic, and molecular levels.

References

[1] Lipson, Hod, and Melba Kurman. *Fabricated: The New World of 3D Printing.* John Wiley & Sons, 2013.

[2] Hiller, Jonathan D., and Hod Lipson. "Design automation for multi-material printing." *Solid Freeform Fabrication Symposium (SFF'09)*, 2009.

[3] Hiller, J., and Lipson, H. "Automatic Design and Manufacture of Soft Robots." *IEEE Transactions on Robotics*, 28(2), 2012, pp. 457–466.

[4] Hanus, Monica J., and Andrew T. Harris. "Nanotechnology innovations for the construction industry." *Progress in Materials Science*, 2013.

[5] Cheney, N., MacCurdy, R., Clune, J., and Lipson, H. "Unshackling Evolution: Evolving Soft Robots with Multiple Materials and a Powerful Generative Encoding." *Proceedings of the Genetic and Evolutionary Computation Conference (GECCO)*, 2013.

Chapter 19

Conclusion: Another Look at Semiautomating the AEC Sector

Alfredo Andia and Thomas Speigelhalter

1910s–1930s: Explosive Industrialization

From the 1910s to the 1930s Europe and the United States saw a period of rapid industrial transformation. An explosive and diverse number of strange machines such as motorized carriages, electric cars, iron-casted train stations, zeppelins, eccentric airplanes, all sorts of air balloons, and an ever-growing number of new industrial apparatuses were popping out every day and everywhere.

1950s: Standardized Industrialization

However, after World War II the imagery of industrialization began to consolidate in an array of advancements. Since the 1950s, cities, infrastructures, and professional services have only transformed gradually in the developed world.

2010s: Explosive Digital Innovation

It is reasonable to argue that today we are well into a new period of accelerating changes in which many inventions in most fields (including the AEC sector) are challenging the industrial processes, products, and services that settled in the 1950s. In this book we put forward several narratives on how this is occurring in a time in which a whole array of inventions such as strange robotic machines, flying cars, algorithmically guided quadricopters, or brain-machine interfaces emerge. The story is not only a technological one but the social construction of technological narratives. The impacts of the digital used to be in the realm of media and work processes but today it is also beginning to alter the way we construct our analog world.

Who Owns Innovation Outside the AEC industry?

Many non-AEC professions have also innovated relatively slowly since the 1950s. For example, traditional health care fields and processes are relatively similar to the ones that consolidated in the medical profession 60 years ago. Yet the branches of medicine that are intertwining with information technology such as DNA sequencing or AI in cancer treatment are growing exponentially as digital technology does. What we are observing today in many industries is that the most radical changes in the professional disciplines usually do not come from the professions themselves but from other disciplines.

Today sectors in many industries are becoming information technology businesses. It is very difficult to find a sector in the economy that has not been impacted. Look at what has happened to sectors like photography, printing, or cinematography in the past 15 years. The car industry is becoming increasingly a digital business with digital manufacturing, the driverless car, and nongasoline-based cars. Similarly, journalism, the oil and gas industries, aerospace, and even the defense industry (with electronic warfare and drones) have been significantly converted into digital businesses. All creating transformative disruption in industries that were considered mature and impossible to break-in. And who owns the innovation? The social units that are able to develop semiautomated narratives.

Platforms of Digital Innovation

Information technology is a chameleon. The computing image is constantly changing, shape-shifting, self-transforming, and disrupting. One way of looking at computing is that it is coming in an array of platforms. These information technology platforms usually began as prototypes. In an initial period they grow slowly but when they pickup momentum they grow explosively. Examples of digital platforms are the mainframe, personal computers, smartphones, the Internet, 3-D printing, robotics, AI, brain-machine interfaces, wearable computers, and many more in the expanding universe of inventions.

AEC Social Units

AEC professionals are consumers of computer technology. Very few members are involved in the basic research but for the most part they are technological clients. The sector is full of relatively small social units that work for particular projects that have very little R&D budget or access to government grants.

How can AEC professionals develop imaginations as these different technological platforms pickup and grow explosively? It was just a decade ago that professional designers began to talk about BIM and parametric CAD. Today, algorithms that semiautomate design and engineering processes are beginning to emerge. Will the information technology innovators of today lead to creating a better analog world?

Today a significant part of the construction industry is still dedicated to designing and assembling industrially manufactured materials and products. Will the AEC sector intertwine with the speed of innovation associated with information technology? From where will the most extreme innovations come in re-creating our analog world?

Digitally Disrupting Platforms in the AEC Industry

There are at least three families of digital platforms that at this moment have the highest potential to transform our conversations about the automation of the AEC sector:

1. Machine learning algorithms;
2. Big data;
3. Digital and biological manufacturing.

Machine Learning: Automating Design

In Chapter 17, we introduced machine learning. We showed a Bayesian network method developed at Stanford that can generate 100,000 iterations of a floor plan in 35 seconds based on algorithms that learn from data extracted from previous plans. There is no human intervention in the fast abundant generation of architectural floor plans. There are many other similar methods surfacing at other university labs in projects such as automating design aspects within engineering, urban planning, social housing, or interior layout [1–3]. In Part II, we also presented a case in which algorithms allowed the architects to design an office building without any heating or cooling in cold Austria. The delicate algorithms optimized the flow of energies within the inside and outside the building.

These methods are still young. They remain semidormant at university labs. They are evolving as peer-reviewed papers led usually by a couple of computer science students and a head professor. As machine learning advances and personal mainframe computing speeds power to magical levels we will probably see a version of this tech platform evolve rapidly at some point.

Big Data: Automating Planning and Real Estate Development

Architectural, engineering, and environmental design synthesis processes will be further be challenged as real-time data from sensors and actuators embedded in all sort of devices continue to explode. Machine learning algorithms are transforming our imagination about how that data can be analyzed and acted upon. One area of impact is real estate development and urban or transportation planning.

Today companies that collects wireless data, such as Airsage, receive over 15 billion anonymous location data points per day in the United States. Using that data, start-ups such as StreetLight Data can do population analytics and predict which groups of people drive and forecast their demographics and their destinations. Small and large companies are using these analytics for retail site selection, marketing, or commercial real estate. City-related groups are using these data analytics for promoting economic development in disadvantaged communities.

Big-data-driven design has been used by many firms to help understand complex capital planning processes for clients in health care, manufacturing, and higher education. Crowd-sourced place making is helping predesign and postoccupancy evaluation by using dynamic data from a large number of users. In-depth postoccupancy and traffic flow evaluations have helped designers understand how users will respond to design decisions in emergency rooms and airport gathering areas.

Big data analytics is allowing energy companies to develop predictive models of energy consumption on a household-per-household basis. It is also allowing planners to observe the impacts of policy decisions with a granularity that never existed. Arup and the Royal Institute of British Architects concluded a joint report that explores big data analytics in urban design [7]. Centers like the MIT Sensible City Lab have already developed a large number of projects that study how we can reshape our cities with data. In short, predictive analytics is rapidly emerging with a family of platforms that will aid in semiautomating the usability, city and infrastructure planning, real estate, and construction markets.

Digital Manufacturing: Automating Construction

As described in detail in Chapter 18, digital manufacturing platforms are emerging in three different stages:

1. First, 3-D digital manufacturing via platforms such as CNC, laser cutters, 3-D printers, and robotic arms allow AEC professionals to manipulate traditional materials at the human scale.

2. Second, 4-D digital manufacturing platforms such as multimaterial printers and nanotechnology can alter the performance of matter at the microscopic, atomic, molecular and supramolecular scale.

3. Third, n-D digital manufacturing via platforms such as programmable matter and synthetic biology will allow us to program matter to change its physical properties at will.

Examples of Emerging Digital Manufacturing: Robotics

Today, we are observing an increasingly larger number of computing platforms in digital manufacturing processes. One example is robotics. There are already more than 1.3 million industrial robots around the world. Yet most of these apparatuses are very inflexible and expensive and work just in very specific manufacturing processes.

However, cheaper robotic arms are rapidly becoming highly versatile and are at the same time dropping in price, making them affordable to a large number of production units and amateurs. DARPA's Autonomous Robotic Manipulation-Hand (ARM-H) grant program is funding the rapid advancement of software and hardware for robotic arms that can perform highly complex tasks. Under that program SRI International has designed a dexterous ARM-H robotic hand that is an electrostatic clutch that if it is mass-produced could cost approximately $3,000—10 times less than preceding technology.

Social adoptions of new computing platforms are very much related to usability, price, and robustness of the platform. Digital platforms that grow disruptively usually explode when they have a stable operating system, which allows a large number of developers to easily create applications for the real world. Think about the smartphone platform. It took the establishment of the iOS and Android operating systems to see an explosive number of companies developing applications from a grassroots level. Similar to the smartphone industry, the robotic industry doesn't yet have a robust robotic operating system (ROS). As Google moves into the robotic business we should expect a maturation of ROS and subsequently a large array of inventions in robotic apparatuses that will take multiple shapes and forms.

Examples of Emerging Digital Manufacturing: 3-D Printing

Like robotics, large-scale 3-D printers are coming in an array of new inventions that are materializing every day. For the construction industry the most critical aspect of the 3-D printing platform is not the printing device but the materials.

In Shanghai, China, Winsun Decorative Design Engineering has 3-D printed several 200 m² (2,000 square feet) houses in 24 hours, each with a reported construction cost of less than $5,000 per house. The gigantic printer is 1.5m (L) x 10m (W) x 6.6m (H), and deposits the material in layers taking into consideration all the MEP systems. The house was printed using recycled concrete, construction waste, cement, and glass fiber. The company has more than 70 patents in new construction materials [4, 5].

Another remarkable project is the next generation CLT—USDA-NIFA FPR Forest Products Research at Washington State University. Researchers Todd Beyreuther, Don Bender, and Dan Dolan developed hybrid cross-laminated timber panel assembly designs and processes to leverage the existing environmental and economic advantages of mass-produced forest residual products, wood reuse, and upcycling for an engineered feedstock. The research positions forest residuals, and reused and recycled construction wood, as a preferred mass-customized building assembly material in sync with emerging parametric design and digital fabrication methods for additive and subtractive manufacturing across the U.S. Northern Rockies Corridor (Figure 19.1) [6].

These 3-D printers are still reimagining materials at the macroscale level. In parallel, in the next decade, we are entering into an era in which microscale 3-D printing can combine diverse types of materials precisely at the nanoscale level, creating all kinds of new materials. Among the many examples emerging in microscale 3-D printing are the ones developed at the Lewis Lab at Harvard University, McAlpine Research Group at Princeton, or the Creative Machine Lab at Cornell University, among many others. As 3-D and 4-D printing platforms enter into a period of explosive growth we might possibly began to observe an explosion of new material patents, faster, and different types of printers, and constant reduction of the cost of construction. This changes will be further accelerated as digital programmable matter and synthetic biology will be able to manipulate the form and performance of materials with processes similar to the ones found in nature.

Routes of Digital Consumption

The discourses of digital innovation cannot be complete if we do not look at the patterns of social adoptions of computing. These patterns are drastically changing. There have been two major narratives of consumption of digital technologies: trickle-down and trickle-up types of information technology consumption.

1. *Trickle-down consumption.* Traditionally, since the 1950s digital platforms have followed a similar maturation process. A computational platform first emerges as an expensive technology that can only be acquired by large organizations. With time, digital technologies become less expensive and more powerful at the same time. In the process, smaller organizations, such as accounting offices and AEC firms, begin to use it to reimagine their business processes. Eventually, these technologies became worldwide consumer products. For example, in the 1950s mainframes could only be afforded by the departments of defense, and banks, and in the 1970s robotics could be consumed only by the largest manufacturing organizations, but in the 1980s most smaller organizations and eventually every citizen had access to computing power.

2. *Trickle-up consumption.* A second route of digital consumption has emerged only in the past decade. In this model digital products are

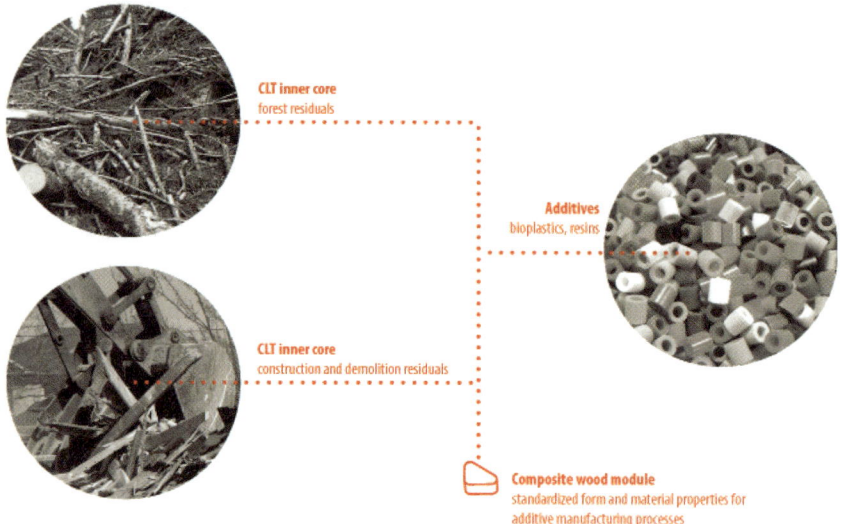

Figure 19.1 Next generation of cross-laminated timber (CTL): Additive processes with engineered feedstock in new urban markets close to rural mills. (Source: Todd Beyreuther, 2014.)

developed from the grassroots. First, the traction of the technology is defined by the consumer who massively uses the technology first (in most cases for free), then smaller organizations, then medium-sized social units, then larger ones, and eventually the government. Crowd-funding, incubators, accelerators, university labs, and social labs have been feeding trickle-up advancements in the past decade. Initially, this model was associated with smartphones and Internet apps, but today in a very short period many new trickle-up products and processes are emerging to disrupt a large number of industries.

We are beginning to live in a different historical moment of digital innovation in which the process of moving from "an idea" to a "billion dollar" company with millions of users can take just a few years by using trickle-up consumption models. For example, WhatsApp moved in 5 years from an idea to a company that process 64 billion messages per day (10 times of the volume of United States' text messages) and was acquired by Facebook for $19 billion when the company had only 55 employees. WhatsApp is now positioning the company to offer voice calls with the aim of disrupting the worldwide wireless industry—transforming an industry with a value of $185 billion in 2012 according to the Cellular Telecommunications Industry Association (CTIA). Other examples of trickle-up companies are Occulus VR, Uber, AirBnb, Space X, Tesla Motors, and a significant number of the emerging Internet companies. Models of trickle-up processes are emerging in the AEC industry via crowd-funding methods to finance real-estate projects via a large number of small investors and crowd-sourcing via platforms such as arcbazar.com that allows clients with small-to medium-scale architectural, interior, and landscape projects to connect with designers in small design competitions.

Cautionary Tale

Both trickle-down and trickle-up modes of consumption of information technology are at play today. The only thing certain is that the social units associated with them, such as tech startups, are aiming to "hack" all industries and human processes. There will be economical booms or busts in the process but there is a clear tendency of continuous acceleration in both digital platforms and innovative social organizations.

Automation has social consequences. Its appearance in everyday media produces both exaggerated enthusiasm and catastrophically dystopian visions for the world. What happens when data is put in a decision-making role? In the book *Automate This: How Algorithms Came to Rule Our World*, Christopher Steiner traces the rise and crash of algorithms in many spheres. He documents the temporary collapse of automated algorithms in book pricing at Amazon, which raised the cost of books into the millions, and the role of algorithms in the flash crash of the stock market on May 6, 2010, in which the Dow Jones lost around 9% of its value within minutes [7, 8].

We are far from fully automating entire processes in the construction industry, but as we document in this book we are well into semiautomating significant parts of the process. It is difficult to write a cautionary tale on automation in particular when we look at the environmental and population challenges we are confronting today in our world. An industrially based construction industry does not have all the answers.

Moonshot Thinking

Will the AEC social units disrupt their industry at the scale that is required? Will the AEC professionals engage in what Google X engineers call "moonshot thinking"? Moonshot thinking is audacious ideas that try to improve things with gains of at least 10 times, in contrast to typical innovation projects that aim at improvements in the 10% range. Thinking with a 10x goal means throwing out completely the old ways of doing things.

So what could be the biggest moonshot thinking the AEC industry could aspire to? The biggest aspiration of the construction industry is to provide dignified, brilliant, and environmentally conscious dwellings for all human activities. This means significantly improved retrofitted facilities, new buildings, and cities. It requires us to rethink how to produce a highly upgraded analog world that is produced with drastically less effort, materials, at much lower cost, and carbon-neutral.

We humans as a species will not fully advance to a higher level until we solve our basic needs of food and shelter. This means thinking at the 10x to 1,000x level. Semiautomation can be a significantly big step to fully understanding that this moonshot idea is possible.

References

[1] Yang, Yong-Liang, Jun Wang, Etienne Vouga, and Peter Wonka. "Urban pattern: Layout design by hierarchical domain splitting." *ACM Transactions on Graphics (TOG)*, 32(6), 2013, Article No. 181.

[2] Bao, Fan, Dong-Ming Yan, Niloy J. Mitra, and Peter Wonk. "Generating and exploring good building layouts." *ACM Transactions on Graphics (TOG)*, 32(4), 2013, Article No. 122.

[3] Leblanc, Luc, Jocelyn Houle, and Pierre Poulin. "Component-based modeling of complete buildings." Proceedings of Graphics Interface 2011. Canadian Human-Computer Communications Society, 2011.

[4] Fung, Esther. "Rapid Construction, China Style: 10 Houses in 24 Hours—Corporate Intelligence." WSJ, Corporate Intelligence RSS. Accessed on 24 Apr. 2014 http://blogs.wsj.com/corporate-intelligence/2014/04/15/how-a-chinese-company-built-10-homes-in-24-hours/?mod=e2fb.

[5] "10 completely 3D printed houses appear in Shanghai, built under a day." 3ders.org. Accessed on 24 Apr. 2014 http://www.3ders.org/articles/20140401-10-completely-3d-printed-houses-appears-in-shanghai-built-in-a-day.html.

[6] Beyreuther, Todd. "Next Generation CLT - Mass-customization of Hybrid Composite Panels." Proceedings of the 102nd ACSA Annual Meeting, Globalizing Architecture/ Flows and Disruptions, Miami Beach, FL, 2013, pp. 866–872.

[7] Steiner, Christopher, and Walter Dixon. *Automate This: How Algorithms Came to Rule Our World.* New York: Portfolio/Penguin, 2012.

[8] RIBA–ARUP. "Designing with data: Shaping our future cities." Accessed on 24 Apr. 2014 http://www.architecture.com/Files/RIBAHoldings/PolicyAndInternationalRelations/Policy/Designingwithdata/.Designingwithdatashapingourfuturecities.pdf.

About the Editors

Alfredo Andia, Ph.D., is an associate professor in the Department of Architecture at Florida International University, Miami. Dr. Andía combines his interest and expertise in architectural design and digital design technologies. He has researched the implementation of digital technologies in more than 140 firms and organizations since the early 1990s in the United States, Japan, and Europe. He has more than 60 papers published on these subjects. He is the founder and coordinator of the Internet Studio initiative. The "iStudio" has became one of the largest experiences in collaborative design studio education in the world and has been featured in significant media pieces by the BBC, Discovery Channel, NHK, *Government Video Magazine*, AP and UPI wire, and many other media outlets around the world. Dr. Andía holds a master's degree from the Graduate School of Design at Harvard University and a Ph.D. from the College of Environmental Design at the University of California, Berkeley.

Thomas Spiegelhalter is codirector of the Structures and Environmental Technologies Lab in the Department of Architecture at Florida International University and teaches sustainability Graduate Studio and Environmental Systems in Architecture. Spiegelhalter has developed numerous solar, zero-fossil-energy and low-energy buildings in Europe and the United States. Many of his completed projects have been published in international anthologies of architecture such as *Architectural Record* magazine (Design Vanguard Award 2003) or in the monograph *Adaptable Technologies—Le tecnologie adattabili nelle architetture di Thomas Spiegelhalter* by Franco Angeli Publisher. He has received 47 honors, prizes, and awards for his work in competitions and for his applied research in carbon-neutral design computation and benchmarking at international, national, and academic levels. He has masters and bachelors degrees in design and architecture from the University of the Arts in Berlin and a bachelor in engineering and architecture from the University of Applied Sciences in Bremen, Germany.

About the Authors

Keith Besserud, AIA, SOM, is the founder and leader of BlackBox, an applied research-oriented computational design resource within the Chicago office of Skidmore, Owings & Merrill. BlackBox studio leads the development and integration of advanced computational concepts within the multidisciplinary design processes in the office of SOM. In particular the group is interested in exploiting various types and sources of data to guide form-finding design processes; to tap into better information and integrate it more effectively into the earliest stages of design. The group applies its interests and skills at multiple scales, ranging from the product scale to the building scale to the urban scale, and has most recently been developing tools and methodologies related to the semantic modeling of cities and the digital integration of urban systems of systems. Keith received his undergraduate degree from the University of Illinois Urbana/Champaign, master of architecture degree from Georgia Tech and a master of engineering degree from the Stevens Institute of Technology.

Clayton Binkley is a structural engineer in Arup's Seattle office. He specializes in the design of complex and atypical structures and is a leader in Arup's design computation community. He works across many engineering disciplines, leveraging computational tools to produce efficient designs and customized analytical tools.

Lucio Blandini, Ph.D., is a civil engineer and architect. He was educated at the Universities of Catania and Bologna in Italy, and took his Ph.D. at the University of Stuttgart in Germany. Lucio Blandini joined the Werner Sobek Group in Stuttgart, Germany, in 2006 and is now a principal and team leader. He is also a licensed engineer for highrise buildings in Dubai/UAE.

Christian Derix is director of Design Futures at WoodsBagot architects. Until 2014, he directed the Computational Design Research group (CDR) of Aedas architects, which he founded in 2004 in London. Design Futures and CDR develop computational simulations for generative and analytical design processes with an emphasis on spatial configurations and human occupation. Derix studied architecture and computation in Italy and the U.K. and has researched and taught the subject at various European universities since 2001, including University of East London, University College London, Milan Polytechnic, Technical University Vienna and as visiting professor at Technical University Munich. Currently he is associate professor at IE University Madrid and visiting professor at the University of Sheffield.

Paul Jeffries is a structural engineer and software developer in Arup's London office. He specializes in computational design and complex geometry and is responsible for Arup Associates' Parametric Design

Unit, an embedded research and development unit that investigates, promotes, and builds computational tools and methodologies to solve design problems. He has worked on a wide variety of geometrically and logistically demanding projects including the ArcelorMittal Orbit sculpture for the London 2012 Olympic Games, King Abdullah Sports City in Saudi Arabia, and Abu Dhabi International Airport. He also teaches algorithmic design and software skills at the Architectural Association Design Research Laboratory in London and blogs about digital design at www.vitruality.com.

Jun Furuse joined Sekisui Chemical Company, Limited, in Tokyo, Japan, on April 1998. Furuse received his master of science and engineering degree at Aoyama Gakuin University in Tokyo in March 1998.

Lars Junghans, is a Professor at Taubman College at the University of Michigan. His research work is focused on building optimization with a comprehensive view to all aspects of the building thermal behavior including passive and active strategies. Further research work includes building automation technology and its potential in reduced green house gas emission. Questions of improved occupant comfort, new ventilation strategies, and easy-to-install technology will be targeted. He graduated from the Swiss Federal Institute of Technology ETH with Ph.D. in building science. Junghans received extensive practical experience in the engineering firm TeamGMI known for its planning of famous architectural projects in Europe. As a collaborator with the famous architecture firm active in sustainable building designs Baumschlager Eberle he was responsible for the energy concept of the Project 22/26, the first office building without active heating, cooling, and ventilation systems.

Abdulmajid Karanouh is a design architect (B.Sc.), computational designer (M.Sc.), façade engineer (M.Sc.), and currently pursuing a Ph.D. in adaptive systems, Abdulmajid leads multidisciplinary project teams specialized in researching and developing context-driven design via algorithmic thinking to design, engineer, and deliver innovative high-performance building systems adaptable to the dynamic context of users and environment, with special emphasis on form, structure, envelope, materials, construction method, and building identity—all fused as one integrated system. Abdulmajid is currently head of Ramboll Innovation Design and Ramboll Façade Engineering in the Middle East.

Masayuki Katano joined Sekisui Chemical Company, Limited, in Tokyo, Japan, on April 1998. He received his B.A. in civil engineering from the Waseda University in Tokyo in March 1998.

Jan Knippers specializes in complex parametrical generated structures for roofs and façades, as well as the use of innovative materials such as glass-fiber reinforced polymers. Since 2000 Prof. Dr. Ing. Jan Knippers has been head of the Institute for Building Structures and Structural Design (itke) at the faculty for architecture and urban design at the University of Stuttgart and is involved in many research projects on fiber-based materials and biomimetics in architecture. He is also partner and cofounder of Knippers Helbig Advanced Engineering with offices in Stuttgart, New York City (since 2009), and Berlin (since 2014). The focus of their work is on efficient structural design for international and architecturally demanding projects. Jan Knippers completed his studies of civil engineering at the Technical University of Berlin in 1992 with the award of a Ph.D.

Achim Menges is a registered architect and professor at the University of Stuttgart in Germany where he is the founding director of the Institute for

Computational Design. Currently he also is visiting professor in architecture at Harvard University's Graduate School of Design. Achim Menges graduated with honors from the AA School of Architecture in London where he subsequently taught as studio master of the Emergent Technologies and Design Graduate Program from 2002 to 2009 and as unit master of Diploma Unit 4 from 2003 to 2006. Achim Menges has published several books on this work and related fields of design research, and he is the author/coauthor of numerous articles and scientific papers. His projects and design research have received many international awards, have been published and exhibited worldwide, and form parts of several renowned museum collections, among others, the permanent collection of the Centre Pompidou in Paris.

Albert Schuster, holds a double diploma in architecture and in mechanical engineering. His particular fields of interest are parametric 3-D-modeling and complex membrane constructions. Albert Schuster joined the Werner Sobek Group in Stuttgart, Germany, in 2001 and was appointed CEO of Werner Sobek Design in 2011.

Werner Sobek is an architect and consulting engineer. Prof. Sobek heads the Institute for Lightweight Structures and Conceptual Design (ILEK) at the University of Stuttgart in Germany. From 2008 until 2014 he was also Mies van der Rohe Professor at the Illinois Institute of Technology in Chicago and guest lecturer at numerous universities in Germany and in Austria, Singapore, and the United States (Harvard). In 1992, Werner Sobek founded the Werner Sobek Group, offering premium consultancy services for architecture, structures, façades, and sustainability. The Werner Sobek Group has offices in Stuttgart, Dubai, Frankfurt, Istanbul, London, Moscow, New York, and Sao Paulo. All its projects are distinguished by high-quality design and sophisticated concepts to minimize the consumption of energy and materials.

Mathew Vola is a structural design engineer. He drives innovation on his projects through smart use of computational design. Working in Australia, Asia, and the Netherlands, Mathew was involved with a number of internationally acclaimed structures like the Airport Link Tunnel in Brisbane, the Singapore Sports Hub in Singapore, and the National Maritime Museum in Tianjin. He is currently leading the Arup Amsterdam buildings team.

Index

A

Abu Dhabi Education Council competition, 49

Adaptive Mashrabiya solar screen, 61–62, 64, 65

Adaptive materials, 205

Al-Bahar Towers
- adaptive Mashrabiya solar screen, 61–62, 65
- adaptive principles optimization, construction, performance manual, 72–74
- algorithm, 65–66
- algorithmic principles, 31
- algorithmic principles for facade and BAS, 59–74
- algorithmic thinking, 64
- BMS control room operator, 74
- communication, 64
- complexity management, 62–64
- computation, 64
- concept and philosophy illustration, 60
- facade, 60
- first set of algorithmic principles, 66–69
- 4-D java scripting, 66–69
- 4-D parametric/BIM model, 70–72
- geometric composition, 60
- geometry optimization, 70–72
- HMI software, 73, 74
- introduction to, 59–60
- key design elements, 60–61
- main form, 60
- main structure, 60
- universal solution, 64
- updating algorithmic principles and software, 74
- Yuanda Europe principles, 73

Architectural, engineering, and construction (AEC)
- automation themes, 21
- digitally disruptive platforms in, 210
- as information technology business, 15
- professionals, 13
- semiautomating the sector, 209–15
- social units, 35, 210

Architectural design, learning algorithms in, 194

Architectural programs, 197

Architecture
- automation themes in, 20
- computational algorithms, 39–45
- enactive, 51–52
- history of parametric in, 21–22

Architecture automation
- machine learning versus, 23
- themes in, 21

Artificial intelligence (AI), 14, 151

223

Artificial intelligence (AI) (continued)
  general AI as third stage of, 198–99
  machine learning as second stage of, 193
  parametric as first stage of, 193
Assembly process, Stuttgart 21 railway station, 96
Australian Green Star Rating, 83
AutoCAD, 85
Automated fabrication and assembly, 163–70
Automatic guided vehicles (AGVs), 148
Automation
  building layout design, 195–97
  defined, 20
  design computation, 118–19
  as evolving computerization theme, 35
  of floor plans, 195
  green, 27–32
  practice, 27
  themes, 21
  See also Construction automation; Design automation
Automation themes
  in AEC industry, 21
  in architecture and engineering, 20, 21
  in construction, 20, 23–24
Autosketch, 85

B

Bayesian network, 195
Bentley Microstation, 80
BEopt, 127
Bespoke learning algorithms, 36
Big data, 211
Biological model, 182–83
Biological organisms, 198

Biomimetic design principles, 183–84
BOM structuring, 166–67
Braced frame, 40–41
Broad Group (China)
  Broad Sustainable Building (BSB), 156
  BSB sustainability, 158
  building automation systems (BAS), 160
  business model, 161
  conclusion, 161
  cost and time, 158–61
  defined, 155
  energy conservation comparison list, 159
  origins, 155–56
  Sky City, 161
  sustainability vision, 156
  T30 Hotel, 157–58
Broad Sustainable Building (BSB)
  defined, 156
  monitoring systems, 161
  Sky City, 161
  sustainability, 158
Building automation systems (BAS)
  characteristics of, 30
  green, 30–31
  intelligent agents (IAs), 30
  T30 Hotel, 160
Building information modeling (BIM), 13, 20
  Al-Bahar Towers, 70–72
  cloud services and, 28
  interoperability of platforms, 28
  movement beyond, 23
  1 Bligh Street, 77, 78
  parametric, 22
  RFI reduction and, 82
  3-D and 4-D, 178–79
Building layout design automation,

195–97
Building optimization, 37–38
  for building energy demand optimization, 121–29
  built example, 128–29
  calculation time improvement, 127–28
  conclusion, 129
  defined, 121
  discrete parameter methods, 122
  future developments in, 127–28
  goal of, 121
  introduction to, 121–22
  optimization tools, 127
  probabilistic optimization methods, 122–24
  robustness and reliability of calculations, 122
  sequential search algorithms, 124–27
  uncertainties, clarification of, 128
  uncertainty diagrams, 125
  usability improvement, 127
  user friendliness, 122
Building performance rating, 27
Building prefabrication, 155–61
Buildings, generic functions of, 53–55

C

Cable-stayed glass facade, 134–38
Catenary models, 49
Cautionary tale, 214–15
Certification systems, 27
Climate surface matrix, 125
Cloud-computing controlled buildings, 30–31
Cloud services
  example, 28
  interoperable, 29

Complexity
  engaging in, 39–45
  managing, Al-Bahar Towers, 62–64
  systems-based, 39
  in urban scale, 44
Computational algorithms
  in architecture and urban design, 39–45
  conclusion, 45
  genetic algorithms, 41–42
  introduction to, 39
  LakeSIM, 44–45
  search algorithms, 40–41
  systems modeling, 42–44
Computational design
  archetypes of space, 56
  ICD/ITKE research pavilion, 184–86
Computational Design Research (CDR), 36, 47, 48, 49, 53
Computer-aided design (CAD), 13, 21–22, 23
Computers
  as autopoietic, self-organizing, and self-learning systems, 192
  history of, 19
  use of, 19
Concept 2226, 126–27
  defined, 128
  energy demand, 129
  exterior, 126
  interior, 126
Configurational representations, 50–51
CONJECT information life-cycle management, 108–9
Construction, 14
  automation themes in, 20
  semiautomated, 14–15
Construction automation, 14
  cases of, 24

Construction automation (continued)
- custom fabrication, 24
- design automation versus, 20
- digital manufacturing in, 212
- manufactured prefabrication, 24
- themes in, 23–24
- via future of digital manufacturing, 24–25
- via n-D digital manufacturing, 201–7

Custom fabrication, 24

Customized algorithmic engineering, 131–39

Customized prefabrication, 171–79
- introduction to, 171
- Miami Valley Hospital, 171–76
- Riverside Methodist Hospital, 176–77

Custom tool development, 116

D

Design automation, 13–14
- construction automation versus, 20
- for residential building, 194–95
- via machine learning algorithms, 191–99

Design computation
- at Arup, 113–19
- automation, 118–19
- case study, 114–16
- custom tool development, 116
- defined, 113
- DesignLink SDK, 116
- introduction to, 113–14
- SALAMANDER, 116–17
- stadium generator (StaG), 117–18

DesignLink SDK, 116

Digital automation manufacturing process, 141

Digital consumption, routes of, 213–14

Digital manufacturing
- in automating construction, 24–25, 212
- conclusion, 206–7
- eras, 201–2
- 4-D manufacturing, 203–5
- n-D manufacturing, 205–6
- robotics, 212
- 3-D manufacturing, 202–3
- 3-D printing, 212–13

E

Enactive architectures, 51–52

Energy concept
- 1 Bligh Street, 79
- Q1 ThyssenKrupp Headquarter, 107

Engineering automation
- machine learning versus, 23
- themes in, 21

Enterprise programs, 21

Enterprise Resource Planning (ERP), 163

Enzo Ferrari Museum, 131–39
- aluminum roof, 138–39
- cable-stayed glass, 134–38
- cable-stayed glass facade, 134–38
- conclusion, 139
- construction site view, 133
- curved roof of, 131
- entrance portal view, 135
- facade, 133
- Gaussian curvature, 138
- geometrical description, 133
- geometrical segmentation, 134
- geometry, 133–34
- Grasshopper data, 136
- illustrated, 132

introduction to, 131–32

scripting work, 138

structural model of roof and facade, 135

sun shading element, 135, 136, 138

workflow diagram, 131

Equimarginal optimization algorithm, 123, 126–27

Euston Station, 53

Explosive industrialization, 209

EZCT Architecture & Design Research, 204

F

Fabrication

custom, 24

robotic, 184–86

Stuttgart 21 railway station, 95–96

Facades

Al-Bahr Towers, 60

double, 79

Enzo Ferrari Museum, 133, 134–38

1 Bligh Street, 79, 82, 84

Q1 ThyssenKrupp Headquarter, 101–2

Factory design, traditional 2-D processes, 145–46

Ferrari Museum, 38

Finite element 3-D modeling, 93–95

Floating Room project, 55, 56

Floor plans

automation of, 195

optimization, 197

Force-intensity method, 40–41

4-D digital manufacturing, 24

4-D java scripting, 67–69

4-D manufacturing

adaptive materials, 205

defined, 201

digitally controlled, 201

nanotechnology, 205

printing new materials, 203–5

Functional House for Frictionless Movement, 52

G

Genetic algorithms (GAs), 41–42

automated system optimization via, 145–53

MATLAB, 127

as probabilistic search technique, 122–23

as robust search technique, 123

GenOpt, 127

Geometric composition, Al-Bahar Towers, 60

Geometry modeling programs, 43–44

Geometry optimization

Al-Bahar Towers, 70–72

Stuttgart 21 railway station, 93

Green automation

Al-Bahar Towers algorithmic principles, 31

building manufacturing, 31–32

cloud-computing controlled buildings, 30–31

conclusion, 32

with human-computer-interface, 30–31

interoperable, automated, carbon-neutral workflows, 28–29

introduction to, 27

SIEMENS total green building automation system, 31

Green building manufacturing automation, 31–32

Green Building Monitor, 109

H

Harmony search algorithm, 122

Heim Automated Parts Pick-Up System (HAPPS)
- application scope of information, 169
- defined, 163
- as durable for alteration, 169
- efficiency and accuracy of, 169–70
- feedback, 169
- illustrated, 166
- intermediates conversion to objects, 168
- outline of, 168
- summary, 170
- *See also* Sekisui Heim

I

ICD/ITKE research pavilion
- biological model, 182–83
- computational design, 184–86
- computerized simulations and optimizations, 182
- defined, 181
- exterior and interior views, 183
- fiber structure, 183
- illustrated, 181
- introduction to, 181–82
- project data, 186–87
- robotic fabrication, 185
- robotic production, 184–86
- robotic production processes diagram, 186
- transfer of biomimetic design principles, 183–84

Industry Foundation Classes (IFCs), 80

Innovation
- digital, explosive, 209
- outside AEC industry, 209–10
- platforms of, 210

Integrated modeling, 42

Integrated product delivery, Q1 ThyssenKrupp Headquarter, 99–110

Intelligent agents (IAs), 30

Islamic geometric composition, 61

J

Java scripting, 67–69

Just-in-time (JIT) prefab construction schedule, 177–78

Just-in-time (JIT) supply chain production, 105

K

KANBAN supermarket system, 105

KASC Sports Hall and Athletics Stadium, 117

Khalifa-bin-Zayed competition, 48

L

LakeSIM, 44–45

Large-format 3-D printers, 202–3

Life-cycle costs (LCCs), 44

"Living kitchen," 207

M

Machine learning
- algorithms versus, 192
- in architectural design, 194
- automating architecture and engineering versus, 23
- in automating design, 211
- automating design via, 191–99
- classification of, 193
- conclusion, 199
- examples outside of AEC industry, 193–94
- hardware, 197–98
- introduction to, 191
- as second stage of AI, 193

Manufactured prefabrication, 24
Manufacturing, green building, 31–32
Marginal utility (MU), 126
Mashrabiya
    adaptive solar screen, 61–62, 64, 65
    in complexity management, 62–64
    incident angle of the sun and, 68
    populated kinetic units, 71
    populating around towers, 66
    units, illustrated, 63
MATLAB, 127
Menu item master (MIM) codes, 167, 168
Mercedes-Benz Museum, 91
Miami Valley Hospital
    delivery process, 173
    design computation, 171
    ergonomic positions for workers, 176
    exterior curtain wall, 171–72
    implementation of prefabrication, 171–76
    initiatives, 171–76
    inpatient rooms, 172–73
    introduction to, 171
    MEP racks, 174–75
    modular and demountable caregiver stations, 172
    prefab modular bathrooms, 172, 173
    temporary pedestrian footbridge, 175–76
Modeling
    finite element 3-D, 93–95
    geometry, 43–44
    integrated, 42
    systems, 42–44
    urban infrastructure systems, 43
Models
    biological, 182–83
    of designer as user, 49–50
    finite element, 94
    meshed, 94
    of occupant as user, 50
Modular Sekisui unit house, 164–65
Moonshot thinking, 215
Multidirectional search algorithm, 122
Multimaterials printers, 203
Multiobjective genetic optimization, 124

N

Nanotechnology, 205
National Maritime Museum of China
    case study, 114–16
    hall structure, 114
    illustrated, 113
    scripted workflow, 115–16
    topological optimization, 115
National September 11 Memorial Museum, 51
NBBJ
    colocating entire design-build team, 177
    conclusion, 179
    defined, 171
    improvements in prefabrication components, 179
    introduction to, 171
    JIT prefab construction schedule, 177–78
    MEP racks, 174–75, 178
    Miami Valley Hospital, 171–76
    prefabrication performance metrics, 176
    prefab shop, 178
    Riverside Methodist Hospital, 176–77

NBBJ (continued)
 3-D and 4-D BIM models, 178–79
N-D manufacturing, 24–25
 dawn of, 201–2
 defined, 201–2
 programmable matter, 205–6
 self-made robots, 206
 synthetic biology, 206
Net-zero energy, 37
Neuromorphic processors, 197–98
Neuro systems, 153

O

1 Bligh Street
 Australian Green Star Rating for, 83
 benchmarking, 85
 BIM conceptual flow diagram, 77
 BIM management flow diagram, 78
 BIM process images, 78
 black water treatment plant, 83
 building key features and systems, 76–78
 CFD simulations, 81
 conceptual site plan, 76
 conclusion, 86
 custom-designed structures and facades, 75–87
 design integration, 83–84
 double facade, 79, 84
 elliptical shape of building, 77
 energy use concept, 79
 facade details and renderings, 82
 facade mock-up, 82
 finite element analysis, 79
 fire service design, 84
 interoperability with contractor and subcontractors, 84–85
 introduction to, 75–76
 mechanical, electrical and plumbing, 84
 mechanical services, 80
 multidisciplinary CAD to BIM collaboration, 79–83
 multiple views, 75
 naturally ventilated atrium, 76, 78
 participants, 86–87
 post-tensioned and reinforced concrete flooring system, 80
 renewable energy, 85
 space conditioning, 79
 structural analysis, 83–84
 water recycling, 85
Optimal truss research, 40
Organic designing, 55–56

P

Parametric
 conceptualization, 21
 as first stage of AI, 193
 history in architecture, 21–22
 movement beyond, 23
 paradigms, 22
 systems limitations, 191
 workflow, 22
Parametric BIM, 22
Parametric formalism, 22
Particle swarm optimization method, 124
Pattern search algorithm, 122
PLM
 advanced features, 148–49
 case study, 149–51
 defined, 146
 digital factory design and operation with, 146–49
 Gantt chart in Tecnomatix Plant

Simulation, 148
parametric production lines with, 146
Plant Simulation, 147, 148
total life-cycle scenarios with, 146
Volkswagen Group case study, 149–51
PLM Tecnomatix, 146
Polish Embassy, 53, 54
Post-parametric automation
era, 23
overview of, 13–15
Prefabrication components, improvements in, 179
Prefabrication performance metrics, 176
Probabilistic optimization methods, 122–24
Product life-cycle management. *See* PLM
Programmable matter, 205–6
Project management, Q1 ThyssenKrupp Headquarter, 107–9

Q
Q1 ThyssenKrupp Headquarter
bar-code controlled JIT supply chain production, 105
certifications, awards, and honors, 109–10
defined, 99
energy concept, 107
facades, 101–2
fire, security, and building control management, 108
introduction to, 99–101
lamella performance angles, 103, 105
material inventory management, 107–8
OpenGL Performer, 100, 101
overall systems topology, 106
participants, 110
project management, 107–9
room automation, 106
schematic workflow diagram, 100
site plan, 99
solar control and thermo facade prototype, 104
spatial 3-D visualization, 101
structural analysis, 105
sun and daylight prototype control system, 101–7
total green building automation, 109
workflow, 104

R
Radio frequency identification (RFID), 105
Real estate development, 211
Renewable energy, 1 Bligh Street, 85
Requests for information (RFIs), 82
Residential building, automated design for, 194–95
Revit Architecture, 80
Riverside Methodist Hospital, 176–77
Robotic fabrication, 181–87
environment illustration, 185
performance of, 184
processes diagram, 186
Robotic operating system, 144
Robotics, 212

S
SALAMANDER, 116–17
Scripting, 37
Search algorithms, 40–41
Sekisui Heim
application scope of HAPPS information, 169

231

Sekisui Heim (continued)
- arrangement and programming of parts, 166
- automated fabrication and assembly, 163–70
- automated workflow, 164
- bill of parts of assembling process, 165
- BOM structuring, 166–67
- defined, 163
- HAPPS, 166, 167–68
- identification by menu, 165
- intermediates conversion to objects, 168
- introduction to, 163–64
- modular Sekisui unit house, 164–65
- part arrangement system, 167–68
- property inheritance from object to parts, 169
- summary, 169–70
- workflow, 165–66
- workflow illustration, 164

Self-made robots, 206

Semiautomated construction, 14–15

Sequential search algorithms
- defined, 124
- process, 125

Siemens digital (self-learning) factors, 145–53

SIEMENS-PLM, 29
- advanced features, 148–49
- case study, 149–51
- defined, 146
- digital factory design and operation with, 146–49
- Gantt chart in Tecnomatix Plant Simulation, 148
- parametric production lines with, 146
- Plant Simulation, 147, 148
- total life-cycle scenarios with, 146
- Volkswagen Group case study, 149–51

SIEMENS total green building automation system, 31

SIMATIC PCS 7, 152

SIMATIC WinCC, 152

Simplex method, 122

Skidmore, Owings, and Merrill (SOM), 35, 41

Sky City, 161

Social units, in AEC industry, 35, 210

Space
- computational archetypes of, 56
- as heuristic organization, 48
- mapping of actions representation, 48

Space-left-over-after-planning (SLOAP), 50

Space planning
- designing organically, 55–56
- enactive architectures, 51–52
- enactive process, 48
- generative process, 48
- generic functions of buildings and, 53–55
- introduction to, 47
- models of designer as user, 49–50
- models of occupant as user, 50
- relational representation, 50–51
- space-left-over-after-planning (SLOAP), 50
- with synthetic user experience, 47–56

Stadium generator (StaG), 117–18

Standardized industrialization, 209

Structural analysis
- 1 Bligh Street, 83–84
- Q1 ThyssenKrupp Headquarter, 105

Stuttgart 21 railway station
- assembly process, 96
- checking of model information, 93–95
- concrete shell thickness, 91
- division line for chalices, 92
- fabrication process, 95–96
- finite element 3-D modeling and automation, 93–95
- low primary energy requirements, 96
- nonlinear analysis, 92–93
- parametric-algorithmic design, 90–92
- participants, 97–98
- passenger comfort, 97
- renderings, 90
- scripting, 95–96
- structural behavior optimization, 92–93
- structural model for chalice, 92
- supporting structure of station hall, 93
- surface model, 96
- as urban large-scale project, 89–90
- zero-energy station, 97

Sun and daylight prototype control system
- defined, 101
- design-to-factory-file process, 104
- illustrated elements, 104
- overall systems topology, 106
- prototype, 104
- room automation, 106
- shading elements, 102

Sustainability
- BSB, 158
- vision, 156

Swarm intelligence, 151

Syntactic representations, 50

Synthetic biology, 206

Synthetic user experience, 47–56

Systems-based complexity, 39

Systems modeling, 42

T

T30 Hotel
- Broad Sustainable Building (BSB), 156
- BSB sustainability, 158
- building automation systems (BAS), 160
- building of, 157–58
- completion, 156
- construction time, 160–61
- cost and time, 158–61
- energy conservation comparison list, 159
- floor modules, 157
- illustrated, 155
- material movement via trucks, 158

Tecnomatix tools, 29

3-D manufacturing
- defined, 201
- digitally controlled, 201
- large-scale, 202–3
- making any form into 3-D, 202
- today, 201

3-D printing, 212–13

3-D printing technology, 32

Trickle-down consumption, 213

Trickle-up consumption, 213–14

V

Volkswagen Group case study
- defined, 149
- simulating learning curves, 151
- Tecnomatix RobCAD, 150

W

Water recycling, 1 Bligh Street, 85

Workflow parametric, 22

Worldwide implemented sustainability, 27